BIOLOGY
FOR A
CHANGING
WORLD

Core Physiology

Michèle Shuster
New Mexico State University

Janet Vigna
Grand Valley State University

Gunjan Sinha

Matthew Tontonoz

W. H. Freeman and Company • New York

Publisher Kate Ahr Parker

Senior Acquisitions Editor Marc Mazzoni

Developmental Editor Andrea Gawrylewski

Associate Director of Marketing Debbie Clare

Managing Editor for First Edition Elaine Palucki, Ph.D.

Senior Media Editor Patrick Shriner

Supplements Editor Amanda Dunning

Assistant Editor Anna Bristow

Managing Editor Philip McCaffrey

Cover and Text Designers Diana Blume, Matthew Ball

Illustration Coordinator Janice Donnola

Artwork Precision Graphics

Photo Editor Christine Buese

Photo Researcher Elyse Rieder

Production Manager Ellen Cash

Composition MPS Limited, a Macmillan Company

Printing and Binding Quad Graphics–Versailles

Library of Congress Control Number: 2011927807

ISBN-13: 978-1-4292-6824-0
ISBN-10: 1-4292-6824-7

Printed in the United States of America

First printing

W. H. Freeman and Company
41 Madison Avenue
New York, NY 10010
Houndmills, Basingstoke RG21 6XS, England
www.whfreeman.com

Brief Contents

Contents

29. Immune System   583

Viral Mysteries After nearly a century, scientists learn what made the 1918 influenza pandemic so deadly   584

30. Plant Physiology   607

Q & A: Plants Plants have evolved a unique set of solutions to nature's challenges   608

Man versus Mountain

Man versus Mountain

Physiology explains a 1996 disaster on Everest

At 1:17 P.M., on May 10, 1996, Jon Krakauer planted one foot in China, the other in Nepal, and stood on the roof of the world. He was at the highest point above earth's sea level that any human has ever reached—short of standing on the moon. Yet he didn't feel like celebrating. It had taken him 6 long weeks to climb to the top of Mount Everest, and now that he was here his toes ached in the sub-zero cold, his breath came in short, painful bursts, and his head pounded from the altitude. It was a struggle just to stay upright. "I cleared the ice from my oxygen mask, hunched a shoulder against the wind, and stared absently at the vast sweep of earth below," wrote Krakauer in an account of his climb for *Outside* magazine later that year.

Krakauer's journey to Everest was a lifetime in the making. While other kids were idolizing astronaut John Glenn and baseball player Sandy Koufax, Krakauer's childhood heroes were Tom Hornbein and Willi Unsoeld—two men from his hometown in Oregon who, in 1963, became the first climbers to scale the daunting western ridge of Everest. As a teenager, Krakauer became a skilled climber, vanquishing many of the world's most difficult peaks, and he dreamed of one day climbing Everest himself. By his mid-twenties, though, he had largely abandoned the idea as a boyhood fantasy. But old dreams die hard.

In 1995, Krakauer was working as a journalist when a call came to shadow an Everest climb and report on it for *Outside* magazine. The 42-year-old writer-adventurer jumped at the chance. He would join a team headed by the celebrated climbing guide Rob Hall, whose company, Adventure Consultants, had successfully put 39 amateur climbers on top of Everest. Reaching the summit himself would mean enduring a weeks-long ascent from Base Camp, giving his body time to adjust to the high altitude.

> **"I cleared the ice from my oxygen mask, hunched a shoulder against the wind, and stared absently at the vast sweep of earth below."**
> —Jon Krakauer

It would also mean risking his life on a daily basis.

The icy tip of Mount Everest sits at 29,035 feet above sea level; cruising altitude of most commercial jetliners is 30,000 feet. A human plucked from sea level and deposited at this altitude would quickly lose consciousness and die. A climber who has spent weeks adjusting to the altitude can function better at the summit, but not very well, and not for very long. Everest is not only the highest place on earth, it is also one of the coldest. At the summit, where windchill temperatures average −53°C (−63°F), freezing to death is a real possibility.

Despite these dangers–or perhaps because of them–a handful of fearless men and women try to climb Everest every season. And every season, some of them don't come back. There are many reasons for these disasters–poor training,

At the summit, where windchill temperatures average −53°C (−63°F), freezing to death is a real possibility.

unforeseen accidents, raw egotism–but among the most important is basic biology: the human body is not equipped to survive at such extreme altitude, and such extreme temperature, for long.

The Body as Machine

Like a car or a computer, a human body is made up of many parts working together in a coordinated fashion. The parts are organized hierarchically, so that smaller components are organized into increasingly larger units, which are themselves organized into more complex systems. The study of all this intricate hardware is known as **anatomy.**

The result of millions of years of evolutionary tinkering, human bodies have an anatomical structure that is impressively well adapted to living in certain environments and performing

ANATOMY
The study of the physical structures that make up an organism.

certain functions. Our species evolved in the hot, flat savannahs of Africa where environmental conditions favored big brains, opposable thumbs, and bipedal posture–as well as the ability to keep cool (Chapter 20). As a result, modern humans excel at grasping a pencil or looking through a microscope; we do less well swimming at the bottom of the ocean or living on mountaintops. Fundamentally, that's because of how we're put together.

For all living things, the smallest anatomical unit is the cell. Human bodies are made up of trillions of cells, each of which can be classified as one of a few hundred different types. Cells, in turn, are organized into **tissues**–layers of specialized cells working together to execute a particular function. Humans and other animals have four different kinds of specialized tissue–epithelial, connective, muscle, and nervous–which carry out specific tasks in the **organs** of which they are a part. The stomach, for example, is an organ composed of the four types of tissue organized into a compartment for churning and digesting food. At the highest level of organization, organs interact chemically and physically as part of **organ systems.** The digestive system, for instance, consists not only of the stomach, but also of other organs, including the esophagus, small intestine, and liver, which all work together to digest and absorb food **(Infographic 25.1).**

TISSUE
An organized collection of a single cell type working to carry out a specific function.

ORGAN
A structure made up of different tissue types working together to carry out a common function.

ORGAN SYSTEM
A set of cooperating organs within the body.

INFOGRAPHIC 25.1

How the Human Body Is Organized

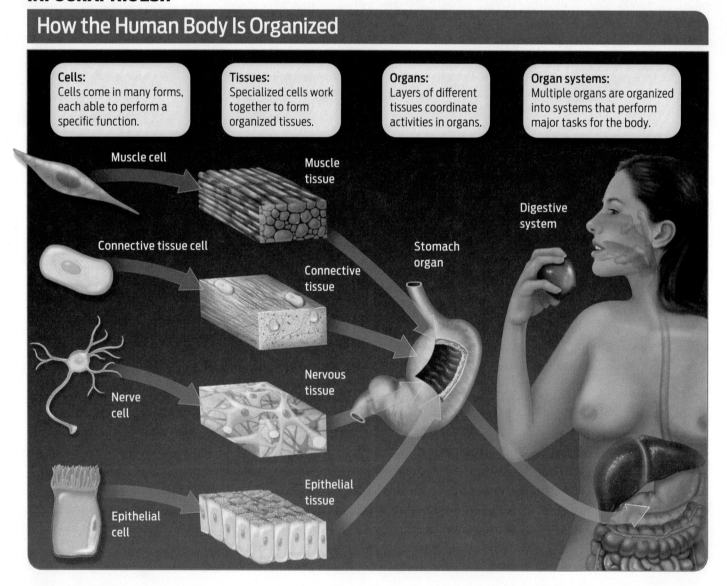

Cells:
Cells come in many forms, each able to perform a specific function.

Tissues:
Specialized cells work together to form organized tissues.

Organs:
Layers of different tissues coordinate activities in organs.

Organ systems:
Multiple organs are organized into systems that perform major tasks for the body.

Muscle cell

Muscle tissue

Connective tissue cell

Connective tissue

Nerve cell

Nervous tissue

Epithelial cell

Epithelial tissue

Stomach organ

Digestive system

PHYSIOLOGY
The study of the way a living organism's physical parts function.

HOMEOSTASIS
The maintenance of a relatively stable internal environment, even when the external environment changes.

THERMOREGULA-TION
The maintenance of a relatively stable internal body temperature.

If the body is like a machine, then physiologists–the scientists who specialize in **physiology,** the study of how a living organism's physical parts function–are interested in how this machine keeps running smoothly. Physiologists want to understand how organ systems cooperate to accomplish basic tasks, such as obtaining energy from food, taking in nutrients to build new molecules during growth and repair, and ridding the body of wastes. To the physiologist, the body is an integrated system for processing inputs and outputs and maintaining **homeostasis,** the maintenance of a relatively stable internal environment.

Thermoregulation: The Physiology of Staying Warm

Like many other animals, humans have an optimal operating temperature and are exquisitely sensitive to temperature changes. Although we

> **Although we can tolerate a wide range of external temperatures in our daily lives, we cannot tolerate even minute changes in our internal temperature.**

can tolerate a wide range of external temperatures in our daily lives, we cannot tolerate even minute changes in our internal temperature. That's because the enzymes that catalyze the chemical reactions in our body function only within a very narrow temperature range (for more on enzymes, see Chapter 4). The body thus works hard to maintain a relatively constant internal temperature compatible with life. Through **thermoregulation,** our body temperature is kept at a consistent–and toasty–37°C (98.6°F).

Keeping a consistent body temperature is just one example of how the body tries to maintain homeostasis. "What we're really thinking about with homeostasis are certain set points that your body needs to maintain," explains Robert Kenefick, a research physiologist with the U.S. Army Research Institute of Environmental Medicine in Natick, Massachusetts. The body has a number of such set points, he says, for things like temperature, blood pH, and blood pressure, and works hard to keep these factors balanced within a very narrow range, even in the face of a changing external environment (**Infographic 25.2**).

Kenefick is an exercise physiologist, and he has spent his career trying to understand how the human body maintains homeostasis during strenuous activities like hiking and running marathons. He works in the Army Research Institute's Thermal and Mountain Medicine Division, where a main focus of his research is understanding how the body performs in extreme cold.

Staying warm is hard to do when ambient temperatures drop below −50°C (−58°F), as they routinely do on Everest. To seal in heat, mountain climbers wear multiple layers of protective gear designed to trap in heat, wick away moisture, and insulate their bodies from the wind and cold. When not hiking, climbers consume copious amounts of hot tea or coffee to warm their insides. But insulated clothes and hot beverages would be of little help if it

INFOGRAPHIC 25.2

The Body Works to Maintain Homeostasis

→ The body expends a great deal of energy to maintain a constant internal environment. Only small fluctuations are tolerated, even in the presence of extreme external conditions.

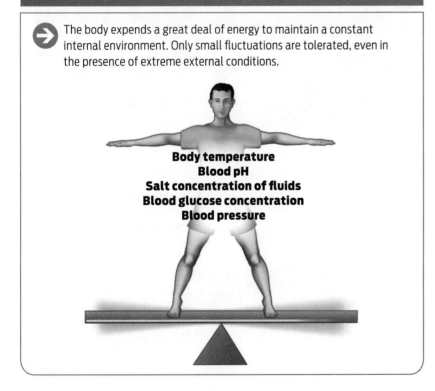

Body temperature
Blood pH
Salt concentration of fluids
Blood glucose concentration
Blood pressure

weren't for the body's natural way of keeping warm.

As Kenefick explains, human bodies respond to cold in two main ways: by conserving the heat they already have and by generating more. To conserve heat, the body performs what is called peripheral **vasoconstriction**–the decrease in diameter of blood vessels just below the surface of the skin. This clamping down of blood flow near the skin surface is why hands, feet, and noses are the first to feel cold on a cold day, and it's a sign that your body is trying to retain heat.

By constricting blood vessels in the skin, peripheral vasoconstriction pushes blood from the skin to the body core, where the internal organs are. "A lot of people believe this is done to increase the amount of blood that goes to your core to help protect those organs," says Kenefick. That's true to a degree, he notes, but the more important reason for peripheral vasoconstriction is "to decrease the amount of heat loss from your skin to the environment."

Like most things in the universe, heat moves along a gradient from higher to lower. "If the temperature is higher in your skin and lower in the air, then you're going to lose heat to the air. By bringing [these temperatures] closer together, you lose much less heat to the environment."

The second way the body responds to cold is by trying to generate more heat. It does this by shivering, which is the involuntary contraction of normally voluntary muscles. "We know that the by-products of cellular respiration–any time cells work, and that includes your muscle cells–are CO_2, heat, and water," says Kenefick. "So when your muscles contract through shivering, they create heat, and that heat helps to warm up the core of your body."

Of course, to maintain a constant temperature, our bodies must not only keep from getting too cold–they must also keep from getting too hot. Two main physiological responses help prevent overheating: peripheral **vasodilation,** the expansion of the diameter of blood vessels, which increases blood flow to the skin; and

evaporative cooling (otherwise known as sweating), which cools the body by releasing heat to the air. In other words, you have a set point for body temperature: if you get too cold, you vasoconstrict and shiver; if you get too hot, you vasodilate and sweat. A precise balance between the two must be maintained to keep tissues healthy. If peripheral vasoconstriction goes on for too long, for example, the result is frostbite–the death of tissues caused by lack of blood flow **(Infographic 25.3)**.

VASOCONSTRICTION
The reduction in diameter of blood vessels, which helps to retain heat.

VASODILATION
The expansion in diameter of blood vessels, which helps to release heat.

INFOGRAPHIC 25.3

Thermoregulation in Response to Cold

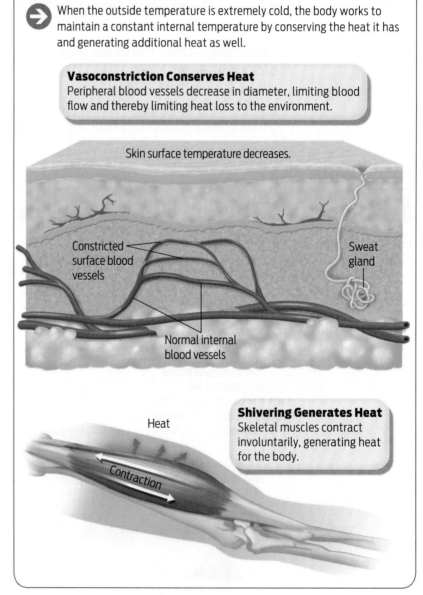

→ When the outside temperature is extremely cold, the body works to maintain a constant internal temperature by conserving the heat it has and generating additional heat as well.

Vasoconstriction Conserves Heat
Peripheral blood vessels decrease in diameter, limiting blood flow and thereby limiting heat loss to the environment.

Skin surface temperature decreases.

Constricted surface blood vessels

Sweat gland

Normal internal blood vessels

Heat

Contraction

Shivering Generates Heat
Skeletal muscles contract involuntarily, generating heat for the body.

Krakauer and his teammates were no strangers to the cold. After weeks of slowly ascending from Base Camp to camps along the route, they reached the launching pad for the summit, the South Col, at 1 P.M. on May 9. "It is one of the coldest, most inhospitable places I have ever been," Krakauer wrote. A wind-swept saddle of rock and ice that sits between the peaks of Everest and neighboring Lhotse, the Col ("col" is a Welsh word meaning "saddle" or "pass") sits at 26,000 feet above sea level. Climbers pitch their tents on the relatively flat terrain and try not to think about the fact that they have entered what's known as the death zone.

Conditions were particularly bad on the Col that day. Gale-force winds blew through the camp, limiting visibility. As Krakauer's teammate, Beck Weathers, later recalled, "The weather was so crummy that when we first got in there, I didn't think there was any chance that we were going to climb that night."

But at 7 P.M., conditions improved markedly. It was still cold—minus 26°C (−15°F)—but the wind had abruptly ceased, and by 11 P.M., above their heads, the stars appeared, while a gibbous moon reflected off the mountain snow. It was the perfect night for a climb.

Into Thin Air

The 15-member team left camp shortly after 11 P.M. Night climbing is necessary in order to arrive at the most difficult parts of the climb during daylight hours and still have enough time to get back down to camp before nightfall. Krakauer led the pack that night, along with the team's head Sherpa, Ang Dorjee.

The pair reached the Southeast Ridge, the penultimate stop along the way to the summit, at 5:30 A.M., just as the sun was peering over the eastern peaks. By this time, Krakauer's hands and feet felt like unwieldy blocks of ice, nearly useless in performing the delicate work

of laying ropes and scaling ice. But it wasn't just the cold he had to deal with. His brain and body were also showing the effects of altitude: "Plodding slowly up the last few steps to the summit," wrote Krakauer in *Into Thin Air,* his 1997 book about the expedition, "I had the sensation of being underwater, of moving at quarter speed."

> ## "Plodding slowly up the last few steps to the summit, I had the sensation of being underwater, of moving at quarter speed."
> –Jon Krakauer

At high altitudes, the percentage of oxygen molecules in the air is the same as at sea level (about 20%), but the barometric pressure of O_2–the number of oxygen molecules banging around in a given volume of the atmosphere–is much lower, as it is with all molecules in air at high altitude. The lower pressure means that fewer oxygen molecules bind to the hemoglobin in blood, which means that blood is less saturated with oxygen, a condition called **hypoxia.** Since all cells require oxygen to function, hypoxia has many bodily consequences. The most serious and immediate occur in the brain. "I've been at 19- and 20,000 feet climbing myself," says physiologist Kenefick, "and I can tell you that doing simple tasks like tying your shoes–even though you've tied your shoes many times–is much more difficult." For the hikers on Everest, he says, each step would have been a struggle.

To help cope with conditions of low oxygen, climbers spend about 6 weeks **acclimatizing** their bodies to the conditions, spending a few nights at progressively higher elevations. Their bodies respond by increasing the output of red blood cells, the cells that contain hemoglobin and carry oxygen. The physiological adjustment of acclimatization allows them to carry more oxygen than could someone coming straight from sea level. But even well-acclimatized hikers usually need bottled oxygen to climb successfully.

At 1:17 P.M., after more than 12 hours of climbing, Krakauer finally reached the summit. It was smaller than he expected–a patch of ice the size of a picnic table, Buddhist prayer flags flapping on a string. He stood and took in the 360° panorama. The towering peaks of the

HYPOXIA
A state of low oxygen concentration in the blood.

ACCLIMATIZATION
The process of physiologically adjusting to an environmental change over a period of time. Acclimatization is generally reversible.

Climbers on Mount Everest, May 1996.

A camp below Mount Everest at night.

surrounding Himalayas were below him, draped in low-lying clouds, like distant swells in a choppy ocean. Beyond the mountain range, endless miles of continent stretched to the horizon, arching slightly with the curve of the earth.

Standing on top of the world, Krakauer was surprised by his own lack of elation. He had just cleared a huge personal hurdle, yet the victory felt hollow. Partly, he was too exhausted to truly care: he hadn't slept soundly in more than 50 hours, and the only food he had been able to choke down in 3 days was a bowl of ramen soup and some peanut M&Ms: sleep disturbances and digestive difficulties are additional side effects of elevation. But another thought lurked in his brain: the oxygen tank he had slung on his back to help him breathe was running low, and he still had to get down the mountain.

"With enough determination, any bloody idiot can get up this hill," guide Rob Hall had famously said. "The trick is to get back down alive." Keenly aware of the clock, Krakauer snapped a few perfunctory photos, and within 5 minutes was headed back down the mountain toward Camp IV.

Fifteen minutes later, after scaling the steep ice fin of the Southeast Ridge, he arrived at the pronounced notch in the mountain known as the South Summit, just below the main peak. As he prepared to rappel over the edge, he caught a glimpse of an alarming sight: a queue of 20 climbers, from three separate expeditions, waiting to come up. They were backed up at the notorious Hillary Step—a 40-foot wall of rock and ice named for Sir Edmund Hillary, who, with Tenzing Norgay, was the first to successfully scale it in 1954. Getting up the Step requires ropes, so climbers must go up one by one, and on this day there was a traffic jam.

While waiting for his turn to get down the Step, Krakauer peered into the distance and saw something he hadn't noticed before: on the horizon, dark clouds were sweeping in from the south, filling up a corner of what had been a clear blue sky. A storm was brewing.

By this point, it was well past the agreed-upon turnaround time of 1 P.M., set by Hall. The climbers who were still headed up the mountain at this hour were willfully flouting safety rules. Not only that, the weather conditions were getting worse. Snow had started to fall, and it had become hard to see where mountain ended and sky began. The lower Krakauer got on the mountain, the worse the weather became.

Krakauer made it back to Camp IV on the Col just before 6 P.M. The bedraggled climber fell into his tent and quickly passed out. He was delirious, shivering uncontrollably, and exhausted. But he was alive.

Sensors Working Overtime

Even as he slept, Krakauer's body was working hard to thermoregulate. Like many physiological processes, thermoregulation is not something that requires conscious thought. It is more like the automated response of a home heating system, triggered when the thermostat is tripped.

The body's thermostat is the **hypothalamus,** a grape-size structure that sits at the base of the brain, right above the brain stem. The hypothalamus receives signals from many different **sensors,** specialized cells in the body that detect changes in both the internal and external environment. For cold, the major sensors are thermoreceptors in the skin and in the hypothalamus itself. Information from various sensors is fed to the hypothalamus, where it is integrated.

Acting as a thermostat, the hypothalamus has a specific temperature set point below which a warning message is triggered that body temperature is dropping. When that happens, the hypothalamus essentially tells the body to take corrective action. For example, it sends a signal to blood vessels in the skin, causing them to constrict in peripheral vasoconstriction. It can also send a signal to muscles to start shivering. Both signals are sent from the hypothalamus to their target tissues along nerve fibers. The cells, tissues, or organs that respond to such signals are known as **effectors:** they act to cause a change in the internal environment. Once the effectors have raised the body temperature, the sensors sense the changed conditions and the signals are turned off.

This circuit of sensing, processing, and responding is an example of a homeostatic

HYPOTHALAMUS
A master coordinator region of the brain responsible for a variety of physiological functions.

SENSOR
A specialized cell that detects specific sensory input like temperature, pressure, or solute concentration.

EFFECTOR
A cell or tissue that acts to exert a response on the basis of information relayed from a sensor.

Homeostasis Feedback Loops Require Sensors and Effectors

➜ By means of sensors, the body constantly monitors factors like body temperature. The sensors relay temperature information to the hypothalamus. If the temperature is too hot or too cold, the hypothalamus sends signals to effector tissues and organs that work to return the temperature to homeostasis levels.

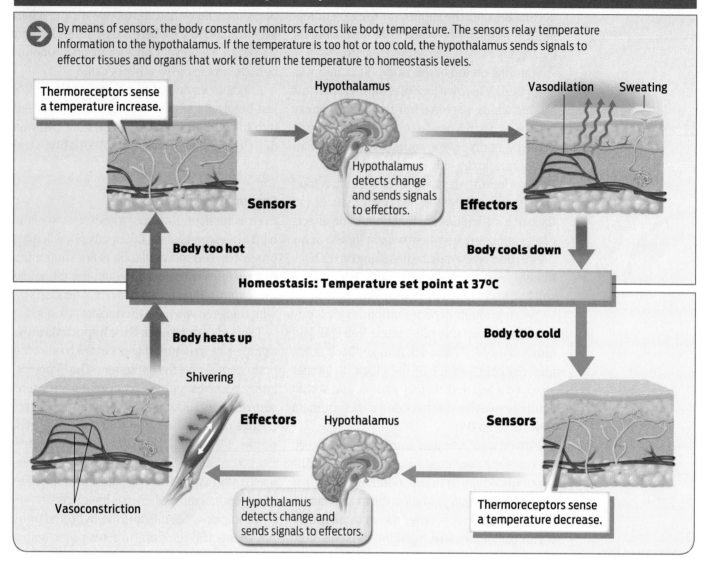

Thermoreceptors sense a temperature increase.

Hypothalamus

Vasodilation   **Sweating**

Hypothalamus detects change and sends signals to effectors.

Sensors   **Effectors**

Body too hot   **Body cools down**

Homeostasis: Temperature set point at 37°C

Body heats up   **Body too cold**

Shivering

Effectors   Hypothalamus   **Sensors**

Vasoconstriction

Hypothalamus detects change and sends signals to effectors.

Thermoreceptors sense a temperature decrease.

feedback loop (Infographic 25.4). In this case, the loop is a negative feedback loop because the output of the circuit (for example, rising body temperature) inhibits the input of the circuit (for example, peripheral vasoconstriction and shivering), thereby helping to bring the system back to its set point. Not all feedback loops act in a negative fashion. Positive feedback loops occur when the output of a system acts to further increase the input of the system. An example is the formation of a blood clot when you cut yourself, which is critical to preventing blood loss. Blood platelets stick to damaged blood vessels

and release molecules that attract even more platelets to the area, which in turn attract even more platelets, and eventually a blood clot forms. Positive feedback loops are effective at rapidly amplifying a response, but negative feedback loops tend to be more common in physiology as they help return the body to its set point and ensure that homeostasis is maintained.

The hypothalamus does more than regulate body temperature. In fact, it is the body's main homeostasis control center, regulating many bodily states including hunger, thirst, and sleep.

FEEDBACK LOOP
A pathway that involves input from a sensor, a response via an effector, and detection of the response by the sensor.

The hypothalamus is part of what Kenefick calls our "lizard brain"—the evolutionarily ancient parts of the brain, which control our most basic physiological responses through unconscious reflexes.

The hypothalamus is able to play such an important role in homeostasis because it is so well connected to sensors and effectors. The hypothalamus is not only a key part of the **nervous system,** connected to parts of the body through nerves, it is also connected intimately to the **endocrine system,** through the pituitary gland. A pea-size structure that sits right below the hypothalamus, the pituitary gland releases **hormones,** which travel through the bloodstream and act on many tissues in the body, including other glands. The endocrine system, with its numerous hormone-secreting glands, is just one of many organ systems found in the human body that cooperate to maintain homeostasis (see **Up Close: Organ Systems** and subsequent chapters in this unit).

During the night, Krakauer was awakened by a teammate who gave him grave news: a number of his teammates, including Hall, had not yet returned to Camp IV. They were still out in the blistering subzero cold somewhere above 26,000 feet. Krakauer's heart sank. He knew the chances of surviving in the cold for that long were slim. By 5 P.M., everyone's oxygen tank would have been empty. It was now midnight. Krakauer feared for the others' lives. But he was also dumbfounded. Hall and the rest of his team were not far behind him on the mountain. What had gone wrong?

The storm began as a cyclone in the Bay of Bengal. It came in low from the valley and then rose up the mountain, gaining in ferocity and strength as it climbed. "One minute, we could look down and we could see the camp below. And the next minute, you couldn't see it," recalled Lou Kasischke, a member of Krakauer's team, who was one of 11 people trapped on the Col when the storm hit and who recounted his

> **"Within the space of five minutes, it changed from really a good day with a little bit of wind to desperate conditions, something I'd never experienced the ferocity of before."**
> —John Taske

experience in the PBS documentary *Storm over Everest.* "Within the space of five minutes, it changed from really a good day with a little bit of wind to desperate conditions, something I'd never experienced the ferocity of before," said John Taske, another member of Krakauer's team, on the same program.

According to Kent Moore, a physics professor at the University of Toronto, the storm that hit Everest that day also caused a particularly severe drop in barometric pressure, greatly reducing the availability of oxygen. "At these altitudes climbers are already at the limits of endurance," says Moore. "The sudden drop in pressure could have driven some of these climbers into severe physiological distress." In particular, they would have experienced the mental side effects of anoxic shock, which include confusion and disorientation.

Unable to tell in which direction they were going, and not wanting to take a wrong turn and step off a cliff, the climbers were forced to hunker down in the hurricane-force winds and wait for the storm to abate. Eventually, after 4 long hours, the clouds parted long enough for one of them to see where they were. Six climbers who were able to walk made it back to camp during this lull. An additional three were brought back safely by the efforts of Anatoli Boukreev, a Russian guide who, having descended to Camp IV, went back to search for them.

But others were not so lucky. Two climbers, too weak to make it back to camp, suffered severe frostbite before being rescued. One lost all his fingers and toes; the other, Beck Weathers, had to have his right hand amputated. Those climbers stuck higher on the mountain—including Hall—could not be rescued. Trapped without shelter in the subzero temperatures all night, their supplemental oxygen and food long gone, the hikers eventually lost their ability to cope with the cold and succumbed to **hypothermia,** a precipitous drop in body

NERVOUS SYSTEM
The collection of organs that sense and respond to information, including the brain, spinal cord, and nerves.

ENDOCRINE SYSTEM
The collection of hormone-secreting glands and organs with hormone-secreting cells.

HORMONE
A chemical signaling molecule that is released by a cell or gland and travels through the bloodstream to exert an effect on target cells.

HYPOTHERMIA
A drop of body temperature below 35°C (95°F), which causes enzyme malfunction and, eventually, death.

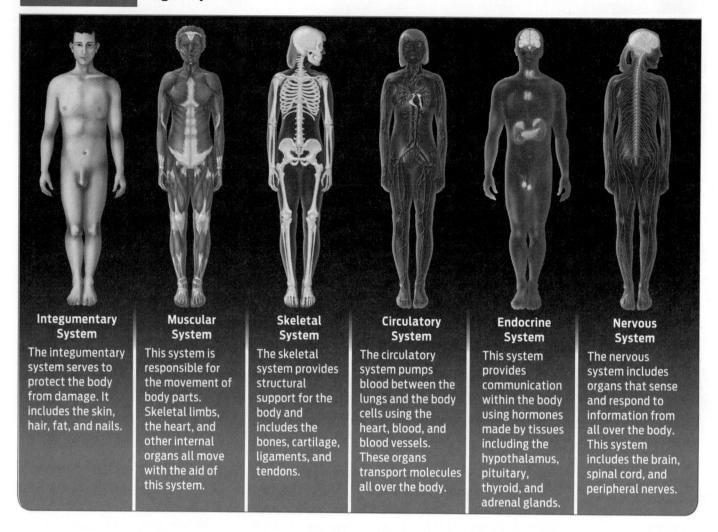

Integumentary System
The integumentary system serves to protect the body from damage. It includes the skin, hair, fat, and nails.

Muscular System
This system is responsible for the movement of body parts. Skeletal limbs, the heart, and other internal organs all move with the aid of this system.

Skeletal System
The skeletal system provides structural support for the body and includes the bones, cartilage, ligaments, and tendons.

Circulatory System
The circulatory system pumps blood between the lungs and the body cells using the heart, blood, and blood vessels. These organs transport molecules all over the body.

Endocrine System
This system provides communication within the body using hormones made by tissues including the hypothalamus, pituitary, thyroid, and adrenal glands.

Nervous System
The nervous system includes organs that sense and respond to information from all over the body. This system includes the brain, spinal cord, and peripheral nerves.

temperature. In all, eight climbers died on Everest that day.

This was not the first time that disaster had struck the summit. A 2008 study of all reported Everest deaths, from 1921 to 2006, led by researchers at Harvard's Massachusetts General Hospital and published in the *British Medical Journal,* found that more than 80% of deaths occurred above 26,000 feet, either during or the day after a summit attempt. While many of these deaths were attributable to traumatic injuries resulting from falls and avalanches, nearly as many were caused by hypoxia and hypothermia.

No Fuel Left to Burn

Although the body is able to cope with cold temperatures for some time through vasoconstriction and shivering, it cannot do so indefinitely. Thermoregulation is work, and work takes energy–roughly 150-300 Calories per hour for a 150-pound man. Eventually, if the body is not consuming food, it will run out of fuel.

The main fuel the body uses in times like this is the sugar glucose, a breakdown product of carbohydrate digestion. When we eat carbohydrates, sugars are released and absorbed into the circulation, and blood sugar increases (see Chapter 4). Some of this sugar may be used immediately as fuel for aerobic respiration in cells of the body (see Chapter 6). Whatever is not needed right away will be converted into **glycogen,** which is stored in muscles and the liver. By converting excess glucose to glycogen, the body maintains a relatively stable blood-glucose level–another example of homeostasis.

Blood-glucose levels are controlled by endocrine tissue in the **pancreas,** an organ that

GLYCOGEN
An energy-storing carbohydrate found in liver and muscle.

PANCREAS
An organ that secretes the hormones insulin and glucagon, as well as digestive enzymes.

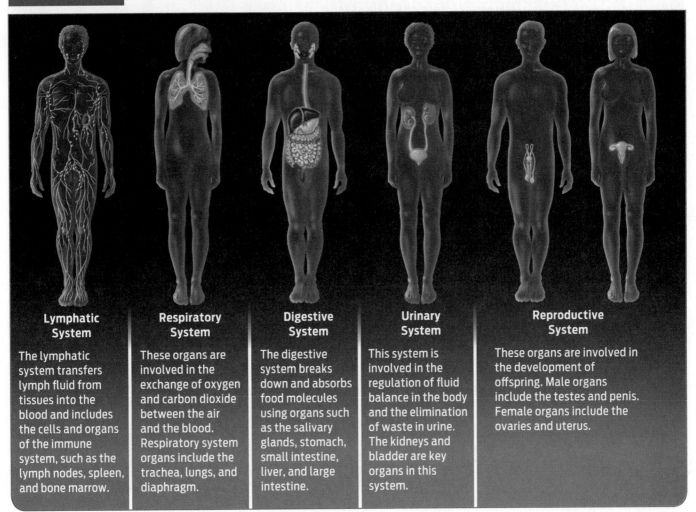

Lymphatic System

The lymphatic system transfers lymph fluid from tissues into the blood and includes the cells and organs of the immune system, such as the lymph nodes, spleen, and bone marrow.

Respiratory System

These organs are involved in the exchange of oxygen and carbon dioxide between the air and the blood. Respiratory system organs include the trachea, lungs, and diaphragm.

Digestive System

The digestive system breaks down and absorbs food molecules using organs such as the salivary glands, stomach, small intestine, liver, and large intestine.

Urinary System

This system is involved in the regulation of fluid balance in the body and the elimination of waste in urine. The kidneys and bladder are key organs in this system.

Reproductive System

These organs are involved in the development of offspring. Male organs include the testes and penis. Female organs include the ovaries and uterus.

INSULIN
A hormone secreted by the pancreas that regulates blood sugar.

GLUCAGON
A hormone produced by the pancreas that causes an increase in blood sugar.

functions in both the endocrine and digestive systems. In response to high blood sugar, the pancreas produces the hormone **insulin,** which binds to receptors on muscle and liver cells, signaling them to remove sugar from the blood. Insulin also signals these cells to make glycogen, using the sugars taken up from the blood.

When blood-sugar levels are low, the body first prompts us to eat by sending a signal to the hypothalamus. If eating isn't an option, the body begins to break down its stored glycogen. The key signal here is **glucagon,** another hormone released by the pancreas in response to low blood sugar, which triggers muscle and liver cells to convert their stored glycogen to glucose. Glucose from the liver is then released into the blood. More glucose in the blood means more energy available to shiver and stay active—and thus warm (Infographic 25.5).

The trapped climbers hadn't eaten in hours, which meant they were operating on glycogen reserves. But the human body can store only so much glycogen. Eventually, after hiking and shivering for many hours, you will exhaust your fuel supply. Without fuel, your body can't continue to remain active and shiver. And if you can't remain active and shiver, then you can't generate heat and your body temperature will fall. That's when hypothermia can set in.

The average adult has enough stored glycogen to power about 12 to 14 hours of routine activity. When a person is exercising strenuously—say, running or hiking—glycogen stores can be depleted in as little as 2 hours. Marathon runners often refer to this point, which occurs at about mile 20, as "hitting the wall." Then, in order to continue exercising, you must eat something—preferably something with carbohydrates.

The Pancreas Regulates Blood Glucose Levels

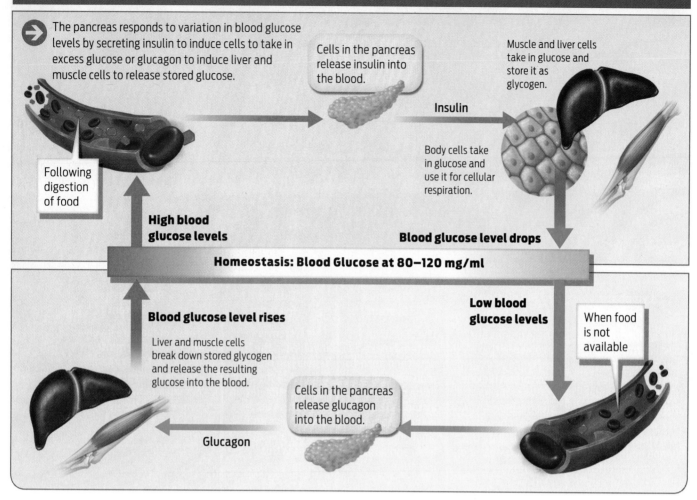

→ The pancreas responds to variation in blood glucose levels by secreting insulin to induce cells to take in excess glucose or glucagon to induce liver and muscle cells to release stored glucose.

Following digestion of food

Cells in the pancreas release insulin into the blood.

Muscle and liver cells take in glucose and store it as glycogen.

Insulin

Body cells take in glucose and use it for cellular respiration.

High blood glucose levels

Blood glucose level drops

Homeostasis: Blood Glucose at 80–120 mg/ml

Blood glucose level rises

Low blood glucose levels

Liver and muscle cells break down stored glycogen and release the resulting glucose into the blood.

When food is not available

Cells in the pancreas release glucagon into the blood.

Glucagon

"A lot of mountaineering communities think you need fat," says Kenefick. "And that's true–fat has more Calories per gram–9 kcals per gram compared to 4 kcals per gram of protein or carbohydrate. But when you're doing things like shivering, those types of contractions, especially, use a lot of glucose." Fats–though a good source of stored energy–are not as readily available for immediate use. And glucose is the primary fuel for the brain.

Fitness also likely played a role in how the Everest climbers fared. Being fit means having more muscle mass than fat. Having more muscle mass means you have more glycogen and can exercise longer and generate more heat through cellular respiration. Someone who is less fit, or who simply has less muscle mass, will tire sooner, need to sit down and rest, and continue to lose heat to the environment. This may be what happened to the climbers who were too weak to hike back to camp: they used up their glycogen stores faster than other climbers.

Another exacerbating factor, says Kenefick, would have been dehydration–a little-known cold-weather risk. In winter, our bodies are working harder under the weight of extra clothing, and sweat evaporates quickly in cold, dry air. We also lose a great deal of water as water vapor when we exhale. The body is about two-thirds water, and when the total water level drops by only a few percentage points, we become dehydrated–which can cause dangerous side effects like delirium, confusion, and convulsions. Kenefick points out that people do not feel as thirsty when it's cold, and thus become even more dehydrated. "We're really tropical animals," says Kenefick. "We came from the Sub-Sahara. We do much better in the heat."

The Kidneys Respond to Changes in Water Balance

The amount of water in the bloodstream controls the concentration of dissolved molecules in the blood and also determines blood volume and blood pressure. The kidneys control water availability by responding to a variety of signals.

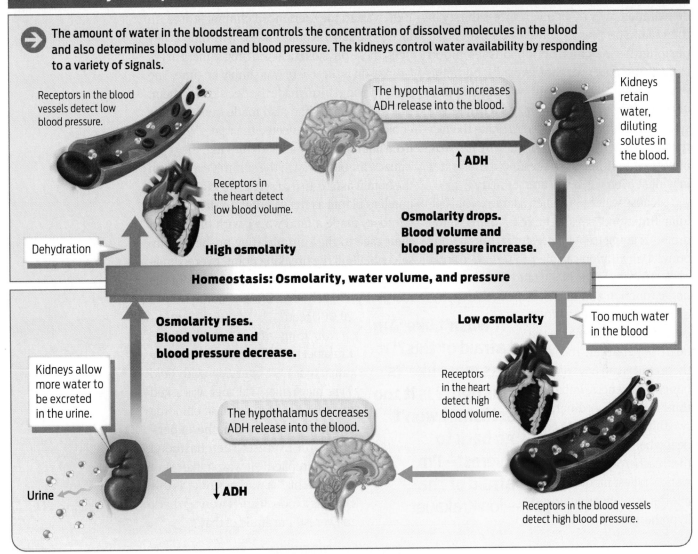

Receptors in the blood vessels detect low blood pressure.

The hypothalamus increases ADH release into the blood.

Kidneys retain water, diluting solutes in the blood.

Receptors in the heart detect low blood volume.

High osmolarity

↑ ADH

Osmolarity drops. Blood volume and blood pressure increase.

Dehydration

Homeostasis: Osmolarity, water volume, and pressure

Osmolarity rises. Blood volume and blood pressure decrease.

Low osmolarity

Too much water in the blood

Kidneys allow more water to be excreted in the urine.

Receptors in the heart detect high blood volume.

The hypothalamus decreases ADH release into the blood.

Urine

↓ ADH

Receptors in the blood vessels detect high blood pressure.

OSMOLARITY
The concentration of dissolved solutes in blood and other bodily fluids.

OSMOREGULATION
The maintenance of relatively stable volume, pressure, and solute concentration of bodily fluids, especially blood.

KIDNEY
An organ involved in osmoregulation, filtration of blood to remove wastes, and production of several important hormones.

As Kenefick explains, our body's sense of thirst relies on **osmolarity,** the concentration of dissolved solutes in the blood. Among the solutes dissolved in blood are electrolytes–ions such as sodium and potassium that are critical for nerve signaling and muscle contraction. Osmolarity is monitored by the hypothalamus as part of **osmoregulation.** When you are dehydrated–when you have less fluid in your blood–the concentration of dissolved solutes is higher. If the hypothalamus registers that the concentration of solutes in the blood is high, it will trigger a sense of thirst, encouraging you to drink. At the same time, it triggers the release of antidiuretic hormone (ADH) from the pituitary, which travels through the bloodstream and acts on the **kidneys.** ADH signals the kidneys to excrete less water in the urine. By reducing the

amount of water lost in urine, ADH causes more water to be reabsorbed by the kidneys back into the bloodstream. Water in the bloodstream dilutes dissolved solutes and lowers the osmolarity. That's why people who are dehydrated have darker urine–it contains less water and is more highly concentrated.

Osmoregulation also depends on sensors that detect changes in blood volume and pressure, both of which depend on the amount of water in the blood. Sensors in the heart, for example, sense how full the heart's chambers are; sensors in blood vessels sense how stretched the vessels are. When low blood volume and pressure are detected, the hypothalamus responds by triggering the release of ADH from the pituitary into the blood, which acts on the kidneys to help retain water **(Infographic 25.6).**

With these multiple sensors for detecting dehydration, why do people feel less thirsty in the cold? The reason, says Kenefick, is that peripheral vasoconstriction pushes blood toward the core. All that extra blood being pushed centrally is sensed by the body as a normal amount of hydration. As a result, the sensation of thirst is reduced, despite the fact that you're dehydrated. This is why it's very important to drink adequate amounts of water in winter, even when you aren't thirsty.

"Because water plays such a large role in cellular function," says Kenefick, "being dehydrated is going to put a greater stress on your body." Dehydration can alter the concentration of electrolytes in the blood, and therefore alter nerve function and muscle contraction. Dehydration also lowers blood pressure and thus makes the heart work harder. Together, these effects can have dangerous consequences, impairing thinking and coordination—two things that matter a great deal when you're navigating the treacherous terrain of the world's tallest mountain during a blizzard, fighting against hypothermia.

> **"It wasn't like 'Am I afraid of this?' It was more like 'Is this right? Is it too selfish?' I won't go back to Everest—I'm afraid of that."**
> —Jon Krakauer

None of the climbers who died on Everest in 1996 was an inexperienced climber–three, in fact, were professional guides. Why didn't they heed these physiological warning signs? Part of the reason is that there was simply no time. The swift-moving storm made the decision for them. But the climbers had also made questionable choices earlier that affected their fate. Whether from hubris or brain-addled thinking, they continued climbing toward the summit even when the hour was late. In that sense, the clock was as much to blame as the weather. In the end, the climbers made a fatal wager with biology: in their race to the summit, they pushed themselves beyond the breaking point, overestimating, in Krakauer's words, "the thinness of the margin by which human life is sustained above 25,000 feet."

Not long after the disaster, Krakauer returned to climbing mountains. In a 1997 interview with *Bold Type* magazine, Krakauer was asked whether he was fearful of climbing again after the trauma he experienced on Everest. The chastened climber replied: "It wasn't like 'Am I afraid of this?' It was more like 'Is this right? Is it too selfish?' I won't go back to Everest–I'm afraid of that."

Warning Signs

Hypothermia isn't only a danger for death-defying mountain climbers: it's a leading cause of death during outdoor recreation like rafting and skiing, and is the number one way to lose your life while outdoors in cold weather. The Centers for Disease Control and Prevention estimate that hypothermia causes more than 1,000 deaths each year in the United States.

Wilderness medicine experts say the best way to prevent hypothermia–in addition to dressing appropriately and carrying plenty of food and water–is to be aware of its signs. In particular, watch for the "umbles": stumbles, mumbles, fumbles, and grumbles, which show changes in motor coordination and altered brain function. If you experience any of these signs in cold conditions, it's time to seek shelter.

Krakauer speaks to journalists on May 16, 1996, after he was flown to Kathmandu, Nepal, from Mount Everest.

How Do Other Organisms Thermoregulate?

In the face of extreme cold, humans strive to maintain a constant body temperature. Although we may pile on warm clothes or sit by a fire, most of our heat is coming from metabolic reactions occurring inside our bodies. Because we use internal metabolic heat to thermoregulate, humans are classified as **endotherms** ("endo," inside). We expend a great deal of energy maintaining a warm body temperature in a cold environment.

Like us, whales are mammals and endotherms. As you can imagine, whales face a great challenge in maintaining a sufficiently warm body temperature in the cold ocean depths. They are protected from this cold by a thick layer of fat tissue called blubber, a type of adipose tissue. Not only is blubber an excellent insulator that helps prevent heat loss to the environment, blubber does not have many surface blood vessels—an adaptation that prevents heat loss from blood to the environment. Many endothermic animals rely on feathers, fur, or fat as insulation.

In contrast to endotherms, other animals must obtain their body heat from the environment and are therefore known as **ectotherms** ("ecto," outside). While these animals are commonly referred to as "cold-blooded," their body temperature actually mimics that of their environment. By using behavioral adaptations, many ectotherms also maintain a relatively stable body temperature (although by definition not as stable as endotherms'). For example, lizards bask in the sun to warm up and seek out shade or a

ENDOTHERM
An animal that can generate body heat internally to maintain its body temperature.

ECTOTHERM
An animal that relies on environmental sources of heat, such as sunlight, to maintain its body temperature.

Insulation Helps Keep Some Endotherms Warm

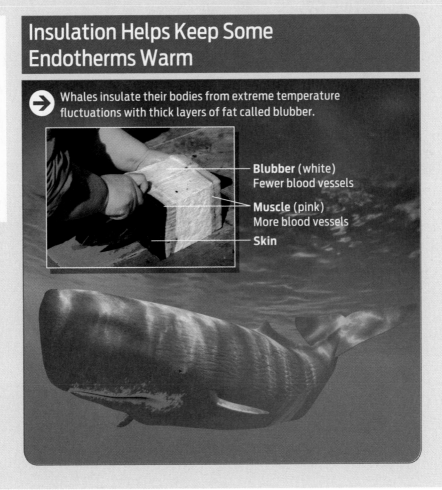

→ Whales insulate their bodies from extreme temperature fluctuations with thick layers of fat called blubber.

Blubber (white)
Fewer blood vessels

Muscle (pink)
More blood vessels

Skin

protected burrow to prevent overheating. Through these behaviors, lizards keep their bodies in a temperature range compatible with their metabolism.

While basking in the sun works well for terrestrial ectotherms, this is not an option for certain fish, such as marlins. The surrounding water absorbs most of the heat of the sun, so marlins cannot heat themselves by swimming in sunny waters. Their brains would be dangerously close to freezing if not for specialized "heater tissue" that generates heat to keep their brains warm. In this case, the heater tissue is modified eye muscle that acts to generate heat rather than force or movement. This form of heating is a type of "nonshivering thermogenesis" ("thermo," heat; "genesis," origin).

Another form of nonshivering thermogenesis, used by bats (and human babies), occurs in a tissue known as brown fat. Brown fat is located in the neck and shoulder areas and is several degrees Celsius warmer than the rest of the body. In brown fat, specialized mitochondria convert energy to heat rather than to ATP, and the many blood vessels in brown fat deliver that heat to other parts of the body.

In the sun, lizards increase their body temperature and become more active.

In the shade, they decrease their body temperature to safe levels.

Some Organisms Generate Heat from Nonshivering Thermogenesis

Ectothermic Marlins Heat Their Brains with Modified Eye Muscle Cells

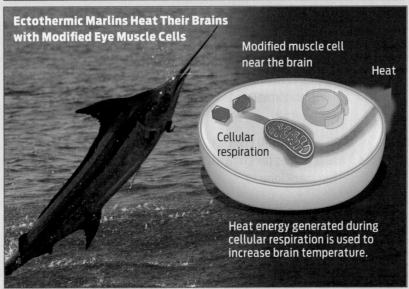

Modified muscle cell near the brain

Heat

Cellular respiration

Heat energy generated during cellular respiration is used to increase brain temperature.

Small Hibernating Endotherms Keep Warm with Brown Fat

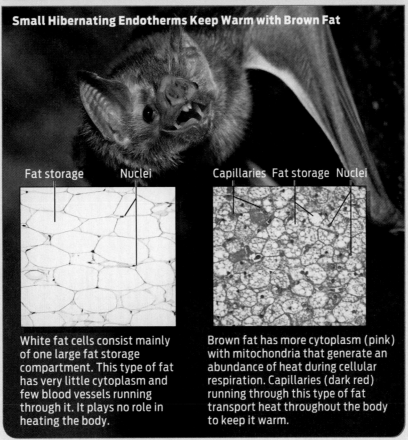

Fat storage   Nuclei

Capillaries Fat storage Nuclei

White fat cells consist mainly of one large fat storage compartment. This type of fat has very little cytoplasm and few blood vessels running through it. It plays no role in heating the body.

Brown fat has more cytoplasm (pink) with mitochondria that generate an abundance of heat during cellular respiration. Capillaries (dark red) running through this type of fat transport heat throughout the body to keep it warm.

▶ Summary

- Living organisms have an anatomical structure that is adapted to suit their physiological functions.

- Humans and other multicellular organisms are organized hierarchically: cells assemble to make up tissues; tissues congregate to form organs; organs work together as part of organ systems.

- Humans have many different organ systems that cooperate to accomplish basic physiological tasks, such as obtaining energy, taking in nutrients to build new molecules during growth and repair, and ridding themselves of wastes.

- Most organisms cannot tolerate wide fluctuations in their internal environment; their bodies work to maintain a stable internal environment, known as homeostasis.

- The process whereby organisms maintain a relatively constant internal temperature is called thermoregulation.

- The body responds to cold temperatures in two main ways: by conserving the heat it has, and by generating more. Vasoconstriction and shivering are key mechanisms of thermoregulation.

- Maintaining homeostasis requires both sensors and effectors. Sensors include nervous system receptors that detect changes in a variety of internal states (for example, temperature and blood pressure). Effectors include the glands or muscles that respond to an abnormal state in an effort to correct it.

- Sensors and effectors work together as part of a circuit or feedback loop.

- Hormones are chemical messengers that travel through the bloodstream, bind to receptors on a target cell, and effect a change in that cell. Insulin and glucagon are hormones that regulate blood-glucose levels.

- Osmoregulation is the control of water balance in the body. Sensors detect blood pressure, blood volume, and solute concentration. Kidneys are important effectors in maintaining water balance.

- Maintaining homeostasis is work and requires adequate energy and oxygen to power cellular respiration.

- Humans (and other mammals) are endotherms: we generate heat internally. Many other organisms, such as reptiles and fish, are ectotherms: they rely on behavior and the environment to maintain a temperature compatible with life.

Chapter 25 Test Your Knowledge

BIOLOGICAL ORGANIZATION

Living things are organized hierarchically: cells are grouped into tissues, tissues into organs, and organs into organ systems.

HINT See Infographic 25.1.

➲ KNOW IT

1. Compare and contrast anatomy and physiology.

2. Organize the following terms on the basis of level of structure, from the simplest (1) to the most complex (4).
___ small intestine
___ mucus-secreting cell of the small intestine
___ digestive system
___ the layer of muscle that contributes to the function of the small intestine

➲ USE IT

3. An emergency room doctor setting a complex bone fracture is relying primarily on knowledge of
   a. anatomy.
   b. physiology.
   c. thermoregulation.
   d. homeostasis.
   e. osmoregulation.

4. Is a personal trainer who works with clients to help them lose weight through a combination of diet and exercise focusing primarily on anatomy or physiology? Explain your answer.

5. Why is the heart considered an organ and not a tissue?

HOMEOSTASIS AND THERMOREGULATION

Most living things work hard to maintain a relatively stable internal environment in the face of a changing external environment. Thermoregulation is the body's way of maintaining a stable internal temperature.

HINT See Infographics 25.2–25.6.

➲ KNOW IT

6. What is homeostasis?

7. Why is glucagon released as part of the response to a drop in body temperature?

8. Describe the feedback loop involved in thermoregulation in cold conditions. Use the following terms in your answer: hypothalamus, sensor, muscle, effector, low body temperature, normal body temperature.

9. People who are severely dehydrated produce _____ urine that is _____.
   a. a high volume of; highly concentrated and dark in color
   b. a high volume of; dilute and light in color
   c. a low volume of; highly concentrated and dark in color
   d. a low volume of; dilute and light in color
   e. a normal volume; a normal color (neither very light nor very dark)

➲ USE IT

10. How could damage to the hypothalamus prevent someone from shivering even if the core body temperature drops dramatically?

11. What conditions might cause high levels of insulin in the circulation? What events would follow?

12. In this chapter you read about homeostatic mechanisms for keeping warm. What responses do you think could help the body dissipate heat during exertion on a hot day? For each mechanism that you propose, give a brief explanation.

13. Tibetan Sherpas, many of whom serve as guides and rescuers on Everest, often do not require bottled oxygen to reach the summit. Why might Tibetans, who have lived at high elevations for many generations, have an easier time than others with hypoxia? (Think about both short-term and long-term changes.)

SCIENCE AND ETHICS

14. What arguments can you make to use tax dollars to pay for basic research into physiology? Refer to some specific examples from the chapter in your answer.

15. The U.S. National Park Service has to rescue stranded hikers, often at great expense. Do you think that hikers' level of preparation should be a factor in determining whether or not they should bear the cost of their rescue? What factors would you consider to determine whether or not a hiker was adequately prepared? Give a physiological reason for each factor that you propose.

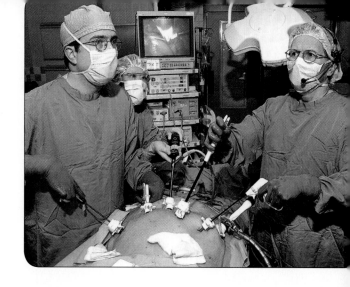

Drastic Measures

Drastic Measures

For the morbidly obese, stomach-shrinking surgery is a last resort

Amy Jo Smith hardly recalls a time growing up when her family wasn't dieting. Her parents were both obese, and they were always trying to lose weight.

Smith herself was relatively slender until her senior year in high school, when her weight began to creep up. She grew up on a horse farm in northeast Maryland. As a teenager she spent much of her spare time on the road, taking her horses to shows. She attributes her weight gain to a diet that consisted primarily of fast food. "I was always eating on the run," says Smith, now 36 years old and a computer literacy teacher. But, she says, her growing girth "never stopped me from doing the things I wanted to do." The extra weight did bother her, though, and she tried several diets and diet pills, only to see her weight yo-yo up and down.

In 2004, at a routine checkup, Smith's doctor noticed that she suffered a number of ills that were likely caused by Smith's 264-pound weight. For one thing, she had been suffering from migraines. She also had stress incontinence—a bladder that leaked when stressed by coughing or laughing, for example. "I thought that was just normal," she says. And she went months at a time without having a period—a telltale sign of a hormonal imbalance often associated with obesity. Smith's physician suggested that, to lose weight, Smith consider having a surgery that would shrink her stomach to the size of a golf ball.

Though we often use the term "fat" casually, obesity is actually a medical condition. A person is considered obese when he or she weighs 20% or more than his or her ideal body weight, based on body mass index. Body mass index (BMI), discussed in Chapter 6, is an estimate of body fat based on a person's height and weight. Morbid obesity—sometimes called clinically severe obesity—is defined as being 100 pounds or more over one's ideal body weight or having

a BMI of 40 or higher. Obesity becomes "morbid" when it significantly increases the risk of one or more obesity-related health conditions or serious diseases. At 5' 2" and with a BMI of 48, Smith was not only obese, she had become morbidly obese.

> **At 5′ 2″ and with a BMI of 48, Smith was not only obese, she had become morbidly obese.**

Even then, Smith had a hard time accepting that she needed such a drastic method to lose weight. "At first I thought he was a quack," she says of her doctor. But she began to think more about the surgery after a friend underwent the stomach-shrinking surgery with stunning results.

There are several types of bariatric, or weight-loss, surgery. ("Bariatric" refers to the study and treatment of obesity; the word is from the Greek *baros,* meaning "heaviness.") The two most common procedures are adjustable gastric banding and gastric bypass ("gastric" means "of the stomach"). In adjustable gastric banding, a surgeon wraps an adjustable band around the stomach to make it smaller so that it holds less food. In gastric bypass, the stomach is surgically made smaller and the small intestine is rerouted. Gastric bypass surgery not only shrinks the size of the stomach but also alters the way food is digested so that the body absorbs fewer Calories. The type of surgery recommended depends on the individual patient's medical history and weight-loss goal. Because there are many associated risks, a National Institutes of Health panel of experts

Smith's weight fluctuated from diet pills and yo-yo dieting.

has recommended surgery only for people considered morbidly obese—people whose risk of death from diabetes or heart disease because of excess weight is five to seven times greater than for those of average weight.

But bariatric surgery is no miracle cure. Because it shrinks the stomach, patients must live the rest of their lives on a strict diet. If they overeat, they suffer nasty side effects such as vomiting and diarrhea.

"It's sort of barbaric," says Monica Skarulis, director of the Metabolic Clinical Research Unit at the National Institutes of Health. Because the surgery so drastically reduces the size of the stomach and restricts how much a person can eat, it amounts to "forced behavior control," she says. On top of that, some gastric bypass patients suffer mineral and vitamin deficiencies over the long term that cause bone loss and potentially other health impairments. And the surgery itself is risky: as many as 20% of patients suffer complications a year after the surgery that are severe enough to put them back in the hospital.

For some morbidly obese people, however, the risk of dying from obesity-related diseases is higher than the risk of surgical complications. And in terms of weight reduction, the surgery is more effective than lifestyle changes alone. Most patients lose 30% to 50% of their excess weight in the first 6 months and 77% after about a year. Studies also show that even 10 years after surgery, most patients still weigh 25% to 30% less than they did before the surgery. Consequently, demand for the surgery is soaring.

Digestion Basics

In Unit 1 we saw that all heterotrophic organisms require food as a source of nutrients and energy, and that to extract both nutrients and energy, our digestive systems break down foods into usable subunits. Here we explore the anatomy and function of the digestive system in more detail, an investigation that will help us understand how a surgery that changes the anatomy of the digestive system can help people lose weight.

Digestion begins immediately following **ingestion**—the act of putting food into our mouths—and consists of both mechanical and chemical processes. These processes occur in the central structure of the digestive system, which is known as the **digestive tract**—essentially a long tube lined with muscles that extends from the mouth to the anus. As the muscles relax and contract, the tube pushes food along. Along its length, this tube receives inputs from various other organs including the salivary glands, gallbladder, liver, and pancreas. The digestive tract's main function is to transform the food we eat into a form our bodies can use. It must also rid the body of the waste left over once usable nutrients and energy are removed from food we have taken in (Infographic 26.1).

DIGESTION
The mechanical and chemical breakdown of food into subunits so that nutrients can be absorbed.

INGESTION
The act of taking food into the mouth.

DIGESTIVE TRACT
The central pathway of the digestive system; a long muscular tube that pushes food between the mouth and the anus.

SALIVARY GLANDS
Glands that secrete enzymes, including salivary amylase, which digests carbohydrates, into the mouth.

INFOGRAPHIC 26.1

The Digestive System

→ The digestive system consists of a long tube with specialized sections and accessory organs that secrete enzymes and other chemicals into the digestive tract. As food travels down the digestive tract, macromolecules are broken into subunits, nutrients are absorbed, and waste is eliminated.

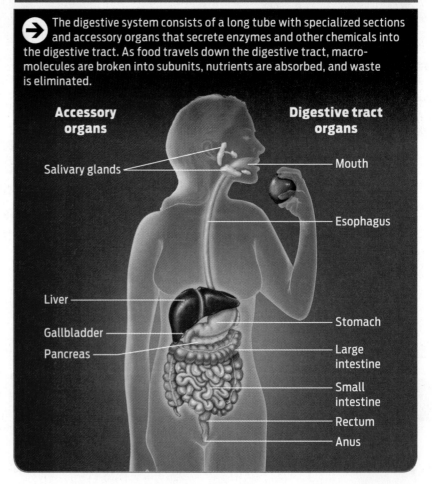

Accessory organs
- Salivary glands
- Liver
- Gallbladder
- Pancreas

Digestive tract organs
- Mouth
- Esophagus
- Stomach
- Large intestine
- Small intestine
- Rectum
- Anus

The Large Intestine

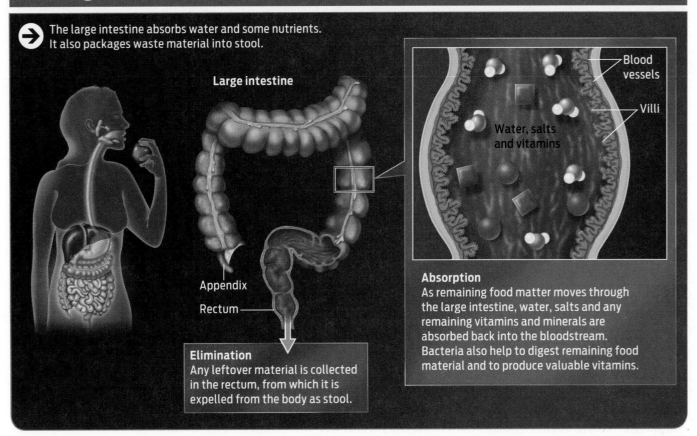

→ The large intestine absorbs water and some nutrients. It also packages waste material into stool.

Large intestine

Blood vessels

Villi

Water, salts and vitamins

Appendix

Rectum

Elimination
Any leftover material is collected in the rectum, from which it is expelled from the body as stool.

Absorption
As remaining food matter moves through the large intestine, water, salts and any remaining vitamins and minerals are absorbed back into the bloodstream. Bacteria also help to digest remaining food material and to produce valuable vitamins.

LARGE INTESTINE
The last organ of the digestive tract, in which remaining water is absorbed and solid stool is formed.

COLON
The first and longest portion of the large intestine; the colon plays an important role in water reabsorption.

STOOL
Solid waste material eliminated from the digestive tract.

ELIMINATION
The expulsion of undigested matter in the form of stool.

the **large intestine,** intact. After chyme passes through the small intestine, it moves on to the large intestine. The large intestine functions like a trash compactor–it holds and compacts material that the body can't use or digest, such as plant fiber. Within the **colon,** the first section of the large intestine, fiber, small amounts of water, vitamins, and other substances mix with mucus and bacteria that normally live in the large intestine. As this waste travels through the colon, most of the water and some vitamins and minerals are reabsorbed into the body through the colon lining. Bacteria chemically break down some of the fiber to produce nutrients for their own survival and also to nourish cells lining the colon (this is one reason fiber is an important dietary nutrient). As the large intestine expands and contracts, it pushes what ultimately becomes **stool** into the rectum, from

which it is **eliminated** through the anus as feces **(Infographic 26.7)**.

Costs and Benefits of Surgery

After looking into various medical centers and attending informational sessions, Smith decided to have her surgery at Christiana Care Hospital in Wilmington, Delaware, in August 2009. She went through the hospital's presurgical screening program, which entailed a number of medical tests and consultation with a team of doctors that included a cardiologist, a pulmonologist, a psychologist, a dietician, and other specialists to ensure that she was physically and mentally fit for the surgery. Often there are undiagnosed medical problems that must be considered or treated before a patient can undergo surgery.

After the surgery Smith lost weight, but it wasn't a smooth ride. The stomach takes time to heal, and so doctors advise patients to ingest only liquids for the first few weeks, puréed foods for the next few weeks, and then gradually progress to solid foods. Because the stomach is now so small, it can carry only about an ounce of food at a time–a handful of crackers or a few broccoli florets. Eating too much at once can cause vomiting or intense stomach pain. But some people find it hard to control their eating habits and wind up in a lot of pain.

Moreover, patients must also stick to a special diet after the surgery or suffer other unpleasant consequences. For example, patients must introduce carbohydrates like breads and pasta into their diet very slowly, says Skarulis, of the National Institutes of Health. If they eat too many simple carbohydrates, the carbs enter the small intestine too quickly. This effect, called gastric dumping, causes nausea and massive diarrhea. But at the same time, patients lose weight.

And research suggests that this weight loss, despite the nasty side effects, does more good than harm: it saves lives. In 2007, Swedish researchers published results from a study in which they followed about 2,000 obese patients who had undergone weight-loss surgery–either gastric bypass or surgical banding–over 15 years and compared them to about 2,000 similarly obese people who didn't have surgery but who were counseled in diet and exercise. After 10 years, those who had gastric bypass surgery weighed 25% less than their presurgery weight; those who had stomach-banding surgery were down about 15%. Those who got traditional diet advice lost no more than 2% of their weight **(Infographic 26.8)**.

More significantly, there were 129 deaths in the diet-only group, mostly from weight-related heart disease and cancer, and 101 deaths in the surgery group–a large difference statistically. Deaths in the surgery

INFOGRAPHIC 26.8

Surgical Procedures and Long-Term Weight Loss

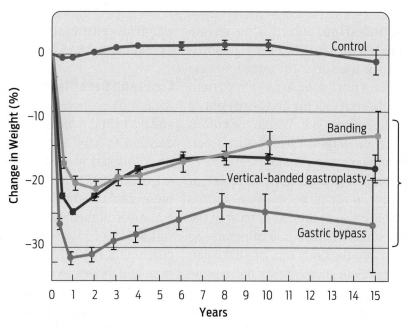

 A 2007 study of obese people showed that those who underwent weight-loss surgery lost significantly more weight than those given only weight-loss counseling.

All forms of surgery resulted in significant weight loss compared with those who did not receive surgery (the control group). The people in the surgery groups maintained significant weight loss over 15 years.

Data represent the mean percent weight change during a 15-year period. Error bars indicate the range of weight loss expected if the experiment were to be repeated.

Source: **Effects of Bariatric Surgery on Mortality in Swedish Obese Subjects** Sjöström et al. *The New England Journal of Medicine*. 2007 August 23; 357 (8): 741–52.

INFOGRAPHIC 26.9

Obesity Surgery Saves Lives

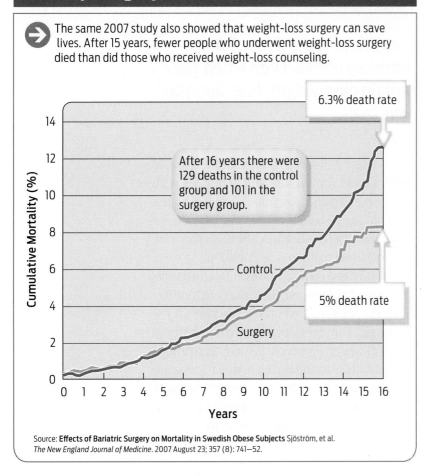

→ The same 2007 study also showed that weight-loss surgery can save lives. After 15 years, fewer people who underwent weight-loss surgery died than did those who received weight-loss counseling.

6.3% death rate

After 16 years there were 129 deaths in the control group and 101 in the surgery group.

5% death rate

Control

Surgery

Cumulative Mortality (%)

Years

Source: **Effects of Bariatric Surgery on Mortality in Swedish Obese Subjects** Sjöström, et al. *The New England Journal of Medicine.* 2007 August 23; 357 (8): 741–52.

group were also mainly from heart disease and cancer, although there were half the number of heart attack deaths in this group compared with the diet group **(Infographic 26.9)**.

Even more encouraging to scientists is the finding that both major types of weight-loss surgery can reverse type 2 diabetes, a condition in which the body can no longer regulate blood-sugar levels and that over time can damage organs and nerves. In 2004, a review in the *Journal of the American Medical Association* of 130 studies of more than 22,000 patients found that 77% of diabetics who have bariatric surgery are cured of diabetes and that 86% are either cured or have their symptoms improve.

No Cure for Obesity

The surgery isn't for everyone. As with any major surgery, there are risks. Although bariatric surgery is much safer today than it was 10 years ago,

1 in 200 patients still dies from the surgery, which can cause complications such as blood clots, hernias, or bowel obstructions. Patients can also end up back in the hospital to repair intestinal leaks that can lead to serious infection.

Smith has had a host of complications that have put her back in the hospital. A few months after her surgery, she felt terrible cramping in her side. Tests showed that scar tissue had formed at the site where her small intestine had been cut from her stomach. Surgery to remove the tissue revealed that part of her intestine and stomach had twisted and anchored onto this scar tissue, which was partly what was causing her pain. Soon after the scar tissue was removed and her stomach and intestines put back in place, she was still having stomach pains after eating. In addition to a feeding tube to her stomach, doctors decided to insert a catheter into a vein in her arm through which she could take nutrients directly into her bloodstream. Smith spent weeks in and out of the hospital between January and April of 2010. But she has had no additional complications that have landed her in the hospital since.

However, her health is still at risk. Since people who have gastric bypass surgery (as opposed to gastric banding) end up with part of the small intestine bypassed, they absorb fewer of the micronutrients they eat. Patients must take such

vitamin supplements as iron, folate, vitamin B_{12}, and calcium for the rest of their lives. There may be additional micronutrient deficiencies that scientists haven't yet recognized; only long-term follow-up of these patients will reveal how serious a problem this is. To monitor her micronutrient levels, Smith has a blood test every 3 months.

What's more, the surgery is not a permanent cure for obesity. As statistics show, most people who have the surgery regain various amounts of weight over time. This is because appetite is controlled by a complicated interaction between the digestive system and the brain. While surgery may reduce the size of the stomach, it doesn't alter the desire to eat, which is controlled by the brain. While scientists are still studying the dynamics of appetite control, they do know that if people do not exercise control over their diet and lifestyle, even those who have had surgery can regain significant amounts of weight.

Although the stomach pouch may stretch over time, it can never be as large as it was before gastric surgery. Most patients never weigh as much as before the surgery. More important, the Swedish study showed for the first time that long-term weight loss for the

morbidly obese, even when they remain overweight, is enough to save lives.

Long-term weight loss for the morbidly obese, even when they remain overweight, is enough to save lives.

The surgery, however, is a drastic measure, as Smith's case shows. She still struggles with nausea every day; strong smells can cause her to vomit. She also feels pain in her left side, for which she takes medication. She also takes anti-anxiety pills at night to help her sleep.

Despite these complications and return visits to the hospital, however, Smith has not "one day of regret," she says. On her first "surgiversary"–her surgical anniversary–she wrote a letter to her surgical team in which she said: "I have been blessed with 35 birthdays but none can compare to my surgiversary. I never imagined in a year that I would lose over 100 lbs, run a 5K the day of my surgiversary . . . sit sideways in a student desk, wear a size 12 pants from a 24-26 . . . be able to sit comfortably in a restaurant booth, and be able to stand on a table or chair without thinking, 'my gosh am I going to break this?'"

By March 2011, Smith had dropped down to 146 pounds. To maintain her weight loss, Smith wakes up at 3:30 every morning to get to the gym to exercise, and she adheres to a strict diet that is high in protein and sparse on simple sugars like sweets. "I don't recognize myself anymore," Smith says.

Of her strict regimen, Smith says, "The surgery is merely a tool. If you aren't willing to make a lifestyle change, it's not going to work for you."

"The surgery is merely a tool. If you aren't willing to make a lifestyle change, it's not going to work for you."
– Amy Jo Smith

How Do Other Organisms Process Food?

Humans and many other animals have what's known as a complete digestive tract–one with two openings, a mouth and an anus. Not all organisms have such a tube-like digestive tract. In fact, many organisms have no digestive tract at all and yet are still able to obtain and process food from their environment.

Take fungi, for example. Fungi are eukaryotic organisms. Some, such as yeasts, are unicellular, while the majority are multicellular. Multicellular fungi have "bodies" that are made up of microscopic filaments called hyphae. However, these "bodies" do not have distinct organs or organ systems, and fungi do not have digestive tracts. To obtain nutrients, fungi extend hyphae into soil (or a piece of bread), where they encounter food such as dead plants and animals. Indi-

vidual hypha cells then release digestive enzymes directly onto their food and absorb the released nutrients directly into their cells. Since digestion occurs outside their bodies, fungi do not need a stomach or a mouth, or even a digestive tract. Rather, each cell can absorb the products of external digestion–digestion that occurs in the environment–directly.

As another example, consider sea anemones, invertebrate animals that live in the oceans, attached to surfaces such as rocks. While sea anemones do digest their food internally, they don't have a digestive tract like humans. Rather, they have a single, multifunctional digestive cavity called a gastrovascular cavity. They can capture food and shove it into this multifunctional digestive cavity using the

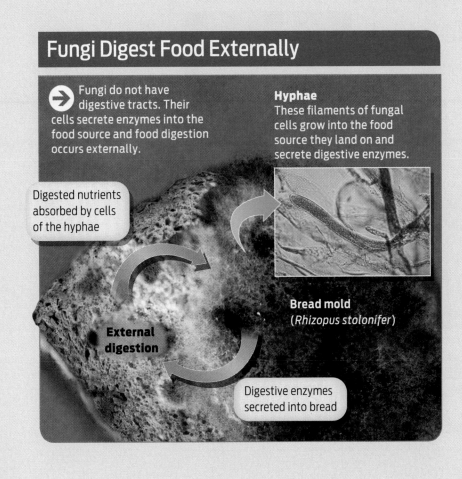

Fungi Digest Food Externally

→ Fungi do not have digestive tracts. Their cells secrete enzymes into the food source and food digestion occurs externally.

Hyphae
These filaments of fungal cells grow into the food source they land on and secrete digestive enzymes.

Digested nutrients absorbed by cells of the hyphae

External digestion

Bread mold
(*Rhizopus stolonifer*)

Digestive enzymes secreted into bread

tentacles that surround their mouths. From the mouth, food enters the gastrovascular cavity, where digestion and absorption of nutrients occur. The gastrovascular cavity has only one opening, the mouth. Food enters through the mouth, and wastes exit through it. This is an example of an incomplete digestive tract, in contrast to our own complete digestive tract in which food flows one way from the mouth to the anus.

Equipped with a stomach or not, all heterotrophic (that is, nonphotosynthetic) organisms need some way of digesting food and absorbing nutrients. As the above examples show, there are many ways to accomplish this task. Photosynthetic organisms such as plants, of course, do not need stomachs or a digestive system because they are autotrophs: they make their own food and therefore don't need to "eat" it (see Chapter 5). Instead, they take up carbon dioxide and use water and the energy of sunlight to convert the atmospheric carbon dioxide to carbohydrates in the plant body.

Sea Anemones Have an Incomplete Digestive Tract

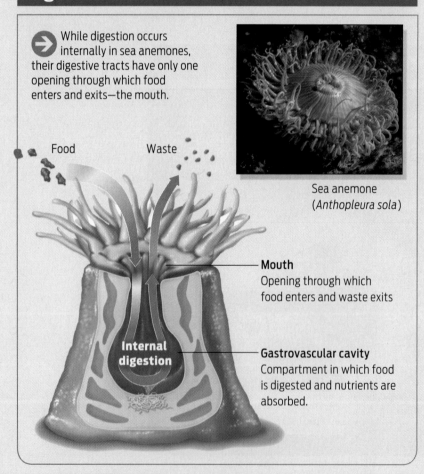

➡ While digestion occurs internally in sea anemones, their digestive tracts have only one opening through which food enters and exits—the mouth.

Sea anemone
(*Anthopleura sola*)

Food

Waste

Internal digestion

Mouth
Opening through which food enters and waste exits

Gastrovascular cavity
Compartment in which food is digested and nutrients are absorbed.

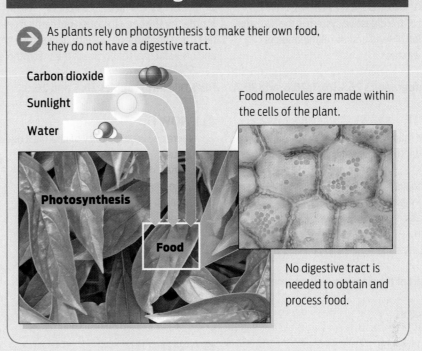

Plants Have No Digestive Tract

As plants rely on photosynthesis to make their own food, they do not have a digestive tract.

Carbon dioxide

Sunlight

Water

Photosynthesis

Food

Food molecules are made within the cells of the plant.

No digestive tract is needed to obtain and process food.

▶ Summary

■ The digestive tract is a long muscular tube that extends from the mouth to the anus and contains specialized organs situated along its length.

■ Digestion begins in the mouth, where teeth chew food (mechanical digestion) and salivary enzymes begin breaking down carbohydrates (chemical digestion).

■ Food passes from the mouth into the stomach through the esophagus, propelled by waves of muscular contractions called peristalsis.

■ The stomach is muscular and acidic and contains a protein-digesting enzyme called pepsin. It is elastic and can expand after a large meal to store food for a few hours.

■ Food processed in the stomach is called chyme. Chyme passes into the small intestine, where enzymes further digest it.

■ Bariatric surgeries reduce the size of the stomach so that it can no longer hold as much food. They may also shorten the small intestine.

■ Enzymes from the pancreas help to digest organic molecules in the small intestine.

■ Bile salts, produced in the liver and stored in the gallbladder, emulsify fats and help the body digest them. Epithelial cells that line the small intestine absorb the broken-down products of food; from there, food molecules enter the bloodstream and are transported throughout the body.

■ The large intestine absorbs water and forms solid stool from indigestible matter in food such as fiber.

■ Humans and many other animals have a complete digestive tract—one with a mouth and an anus. Not all organisms have a complete digestive tract; many have no digestive tract at all.

DIGESTIVE SYSTEM ANATOMY

The function of the digestive system is the digestion and absorption of food and the elimination of indigestible wastes.

HINT See Infographics 26.1–26.3, 26.6, and 26.7.

➲ KNOW IT

1. Place the following structures of the digestive system in order (from the entry of food to the exit of waste).

___ esophagus
___ large intestine
___ stomach
___ mouth
___ small intestine

2. Which part of the digestive tract has the most acidic pH?
   a. esophagus
   b. colon
   c. small intestine
   d. stomach
   e. mouth

3. Why is it helpful to have an expandable stomach?

➲ USE IT

4. What do the gallbladder, liver, and pancreas have in common with respect to the digestive system? How do they differ from the mouth, stomach, and small intestine?

5. Gastric bypass surgery causes the _____ to become _____.
   a. stomach; smaller
   b. small intestine; larger
   c. stomach; less acidic
   d. small intestine; less acidic
   e. stomach; larger

6. Muscle paralysis in the digestive tract would compromise which digestive function?
   a. digestion in the stomach
   b. digestion in the small intestine
   c. absorption in the small intestine
   d. digestion in the mouth
   e. movement of food

DIGESTIVE PROCESSES

In order for the body to obtain nutrients from the diet, food encounters a variety of digestive enzymes and other factors that process the macromolecules.

HINT See Infographics 26.4 and 26.5.

➲ KNOW IT

7. Where does the majority of chemical digestion take place?
   a. small intestine
   b. esophagus
   c. mouth
   d. stomach
   e. colon

8. What do pepsin and salivary amylase have in common?

9. Which organ produces lipase?

➲ USE IT

10. A person who has had his or her gallbladder surgically removed will have trouble processing
   a. fats.
   b. carbohydrates.
   c. minerals.
   d. vitamins.
   e. proteins.

11. Compare and contrast the roles of bile salts and lipase.

12. Why would someone with a blocked duct between the pancreas and the small intestine experience pancreatic inflammation (pancreatitis)? Note that in this case inflammation is a response to tissue damage.

13. If you stand on your head, can processed food still pass from your small intestine into your large intestine? Explain your answer.

14. Why do both people who have had their gallbladders removed and people who take Alli experience "greasy" diarrhea if they eat a high-fat meal?

SCIENCE AND SOCIETY

15. What measures can a person with a very high BMI take to reduce the risk of health complications? If a person with a very high BMI chooses not to alter his or her diet and lifestyle, or is unsuccessful in the attempt to cut the risk of serious medical conditions, do you think that public or private health insurance should cover the cost of treating such a condition? Explain your answer. Consider societal, personal, economic, and genetic circumstances that can contribute to a high BMI.

16. From what you have read about gastric bypass surgery, what would you tell someone who is morbidly obese and who is considering this surgery about its known risks, benefits, and any "unknowns"? Would you say the same to someone considering the surgery who is simply overweight, not morbidly obese? Explain your answer.

SMOKING AREA

Smoke on the Brain

Smoke on the Brain

Nicotine and other drugs of abuse alter the brain and are hard to kick

Jack Ward thought he could resist picking up a cigarette, but the temptation was too strong. He had happily quit smoking in 2006. But when he walked into a smoke-filled poker room 3 years later at a friend's house in Brooklyn, New York, the scene before him seemed to run in slow motion. He watched intently as smokers sucked languidly on their cigarettes, deeply drawing in each puff and then exhaling with sighs of contentment. He resisted that night. But poker night became a weekly event, and finally he gave in: he began smoking again.

Ward knew that smoking is risky. Cigarette smoke is associated with various health problems. He had had countless arguments with his wife, who had insisted he stop smoking to protect his own health and hers as well, which is partly why he quit in the first place. But during poker night, none of that seemed to matter.

Cigarette smoking is highly addictive—smokers can develop a physical and psychological need to smoke. And while anyone might be able to smoke one cigarette or even several and not become addicted, most people find it extremely difficult to stop if they have smoked for an extended length of time. Ward had started smoking casually in high school. By the time he went to graduate school, he was smoking a pack and a half a day. "I was surrounded by people who smoked," Ward recalls. "It never seemed unusual to smoke so much."

Brain scientists have long known that cigarettes, caffeine, and other drugs of abuse such as cocaine, heroin, alcohol, and marijuana stimulate the brain's reward system, a complex circuit of brain cells that evolved to make us feel good after eating or having sex–activities we must engage in if we are to survive and pass along our genes. Without a feeling of pleasure from these activities that ensure our survival, we might never seek out sex or food, and our species would die out. Drugs of abuse stimulate the very same pleasure pathways, which is why

14. Is more or less of the neurotransmitter acetylcholine released by the axon terminals of neurons in patients with multiple sclerosis compared to those in people who do not have multiple sclerosis? Explain your answer.

ADDICTION

Addiction is associated with alterations in brain chemistry and in the levels of certain neurotransmitters and receptors.

HINT **See Infographic 27.7.**

⊘ KNOW IT

15. Addictive substances confer a sense of pleasure because they

    a. decrease the amount of dopamine in synaptic clefts.

    b. increase the amount of dopamine in synaptic clefts.

    c. increase the number of dopamine receptors on the axon terminals of cells that release dopamine.

    d. increase the number of dopamine receptors on dendrites of cells that release dopamine.

    e. c and d

16. Why do drug users need to take ever-increasing amounts of drugs to get the same high?

⊘ USE IT

17. Would you expect a person born with a relatively low number of dopamine receptors to be happier or sadder than the average? Explain your answer.

18. Parkinson disease is caused primarily by a gradual loss of dopamine-producing neurons in the brain. Why are mood swings often among the debilitating symptoms of Parkinson disease?

SCIENCE AND SOCIETY

19. Given that many addictive substances and behaviors (for example, the nicotine in cigarettes; cocaine; gambling) act at least partially through the same biological mechanism—dopamine pathways in the brain—why do you think that each is regulated differently by the government?

20. If you were a researcher at the NIDA, how would you explain some of the broader impacts of your work to a group of taxpayers who, as one of them put it, don't understand "why my tax dollars are going to help a bunch of junkies"?

Too Many Multiples?

Too Many Multiples?

The birth of octuplets raises questions about the fertility business

The live birth of octuplets is an extremely rare occurrence, which has happened only twice in U.S. history. So the arrival of the second set in a California hospital in January 2009 was greeted with fanfare. Headlines read "Octuplets Stun Doctors" and "Eight Babies!"

But days after news of the miracle multiple birth spread worldwide, the public reception turned sour when it came to light that the 33-year-old mother, Nadya Suleman, already had six children all under the age of 7 who were, like the octuplets, born through a process of assisted reproduction called **in vitro fertilization (IVF).**

Even more disturbing to some, Suleman was an unemployed single mother on welfare. The public outcry was fierce. How could she support her children? Was she psychologically disturbed? And why had her doctor agreed to give her another round of IVF when she already had six children?

There were many more questions than answers. And far more important, the case cast a spotlight on the business of fertility treatment: in the United States fertility clinics are unregulated. Although the American Society of Reproductive Medicine (ASRM) issues guidelines on how doctors should administer fertility services, in most states there are no laws regulating what doctors can and cannot do in this regard. Though critics called Suleman's doctor irresponsible, he had not violated any laws.

Infertility treatment wouldn't be nearly as controversial if it didn't increase the odds that a woman would conceive more than one child during the treatment. "Multiples," as these babies are called, are often born prematurely. Consequently they are underweight and have underdeveloped organs, and so are at risk for birth defects such as cerebral palsy and for health problems later in life. Carrying multiples also increases the risk that the mother will develop dangerously high blood pressure, diabetes,

IN VITRO FERTILIZATION (IVF)
A form of assisted reproduction in which eggs and sperm are brought together outside the body and the resulting embryos are inserted into a woman's uterus.

Thanks to in vitro fertilization, Suleman has 14 children.

vitamin deficiencies, or one or another of several other medical conditions during her pregnancy that may affect the health of her unborn children. Even women who have twins have a higher risk of medical complications during pregnancy, and their babies are at a higher risk of premature birth than singletons. Suleman's eight children are the only surviving set of octuplets ever.

Multiple births without fertility treatment are extremely rare. Scientists estimate that the incidence of natural triplets is 1 in 6,000 to 8,000 births—of quadruplets, 1 in 500,000 births. The incidence of higher-order multiple births is even rarer. But because assisted

Scientists estimate that the incidence of natural triplets is 1 in 6,000 to 8,000 births—of quadruplets, 1 in 500,000 births.

reproduction procedures have skyrocketed, so, too, have the number of multiple births. In 2006, for example, the number of triplet births was 12 times higher than normal, according to statistics compiled by the U.S. Centers for Disease Control and Prevention. The medical complications associated with such pregnancies place a large burden on the health care system.

Although there have been efforts to improve assisted reproductive technology and thus reduce the likelihood of multiple births, extreme cases such as Suleman's have drawn lawmakers to the scene. Many states are considering legislation that would place restrictions on fertility doctors. Working from another perspective, some groups are fighting for legislation that would require health insurance providers to pay for fertility treatments. Because many patients pay out of pocket, patients with limited financial resources can, because of their circumstances, pressure doctors to be aggressive with treatment, despite the health risks associated with a multiples pregnancy. The goal of this legislation is to make assisted reproduction safer for everyone, not just for those who can afford it.

The Anatomy of Sex

Up to 15% of couples have trouble becoming pregnant–perhaps a surprising statistic, considering that many women spend years trying *not* to conceive. And most couples are unaware that

> **Successful pregnancies require properly functioning sexual anatomy, balanced and precise hormone secretion, and some lucky timing.**

they have a fertility problem until they begin trying to have a child and nothing happens.

While having children is a natural part of life, the actual process of reproduction is a highly complicated affair. Successful pregnancies require properly functioning sexual anatomy, balanced and precise hormone secretion, and some lucky timing.

Suleman claimed she suffered from a medical condition that prevented her from conceiving a child naturally. If she did, she wasn't alone. In the United States, an estimated one out of eight couples experiences infertility, which is defined as the inability to conceive within a year or to bring a pregnancy to term. Many things can cause infertility: advanced age, infections, hormonal imbalances, chromosomal abnormalities, and physical blockage of reproductive passages.

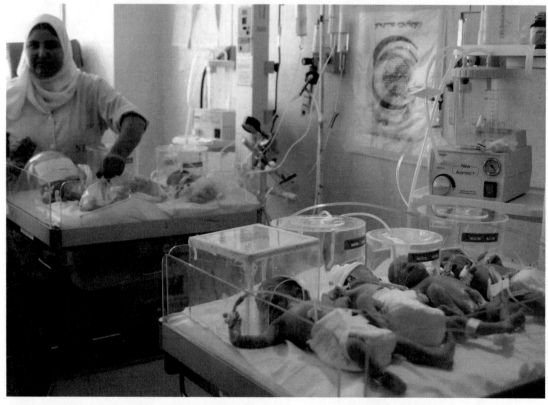

A nurse monitors the newborn septuplets of a woman who gave birth to four males and three females at the hospital of the University of Alexandria, Egypt, in 2008.

OVARIES
Paired female reproductive organs; the ovaries contain eggs and produce sex hormones.

ESTROGEN
A female sex hormone produced by the ovaries.

PROGESTERONE
A female sex hormone produced by endocrine cells in the ovaries (particularly in the cells of the corpus luteum) that prepares and maintains the uterus for pregnancy.

OVIDUCT
The tube connecting an ovary and the uterus in females. Eggs are ovulated into and fertilized within the oviducts.

UTERUS
The muscular organ in females in which a fetus develops.

ENDOMETRIUM
The lining of the uterus.

TESTES (SINGULAR: TESTIS)
Paired male reproductive organs, which contain sperm and produce androgens (primarily testosterone).

SCROTUM
The sac in which the testes are located.

TESTOSTERONE
The primary male sex hormone, which stimulates the development of masculine features and plays a key role in sperm development.

ANDROGEN
A class of sex hormones, including testosterone, that is present in higher levels in men and causes male-associated traits like deep voice, growth of facial hair, and defined musculature.

SEMINIFEROUS TUBULES
Coiled structures that constitute the bulk of the testes and in which sperm develop.

EPIDIDYMIS
Tubes in which sperm mature and are stored before ejaculation.

VAS DEFERENS
Paired tubes that carry sperm from the testes to the urethra.

URETHRA
The passageway through the penis, shared by the reproductive and urinary tracts.

SEMEN
The mixture of fluid and sperm that is ejaculated from the penis.

Men can suffer from a low sperm count or have abnormal sperm. In many cases, the reason for a couple's infertility remains unknown.

According to the ASRM, modern medicine can offer treatment to 90% of infertile couples. But fertility isn't an exact science, and treatment isn't always effective. There are many organs and hormones involved in human reproduction, and communication among them is a highly orchestrated process.

The female reproductive system consists of the **ovaries**–paired structures that hold eggs in various stages of development and that produce the sex hormones **estrogen** and **progesterone,** the major female sex hormones. Each month, one ovary typically releases one egg into an adjacent organ called the **oviduct** (also known as the fallopian tube). Eggs travel via the oviducts to the **uterus,** an elastic muscular compartment that can support a growing fetus. The uterus is lined with tissue called the **endometrium** (Infographic 28.1).

In males, paired glands called **testes** (or testicles) are contained in a sac of skin called the **scrotum.** The testes produce **testosterone,** the primary male sex hormone. Testosterone is one of the **androgens**–a class of hormones that are present in higher amounts in males than in females. Each testis contains tightly coiled **seminiferous tubules** within which sperm develop. Remarkably, each testis contains approximately 250 meters of seminiferous tubules. Sperm travel through the seminiferous tubules and enter the **epididymis,** where they mature and are stored until ejaculated. From the epididymis, sperm travel through paired tubes called the **vas deferens** and exit the body through the **urethra.** Along the way, the male reproductive system adds additional fluid to sperm to make the sperm hearty enough to survive in the female reproductive tract. Ejaculated sperm and the accompanying fluid are called **semen** (Infographic 28.2).

INFOGRAPHIC 28.1

Female Reproductive System

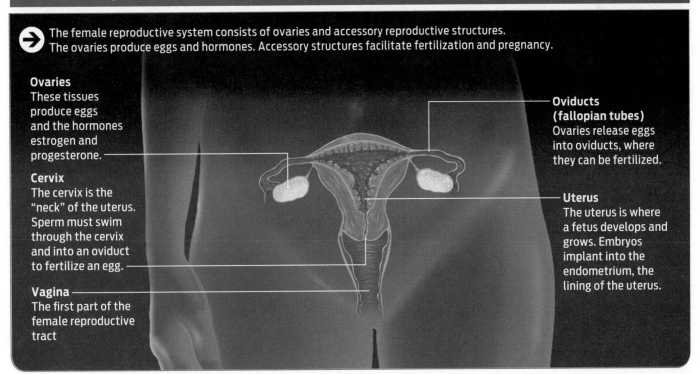

→ The female reproductive system consists of ovaries and accessory reproductive structures. The ovaries produce eggs and hormones. Accessory structures facilitate fertilization and pregnancy.

Ovaries
These tissues produce eggs and the hormones estrogen and progesterone.

Cervix
The cervix is the "neck" of the uterus. Sperm must swim through the cervix and into an oviduct to fertilize an egg.

Vagina
The first part of the female reproductive tract

Oviducts (fallopian tubes)
Ovaries release eggs into oviducts, where they can be fertilized.

Uterus
The uterus is where a fetus develops and grows. Embryos implant into the endometrium, the lining of the uterus.

Male Reproductive System

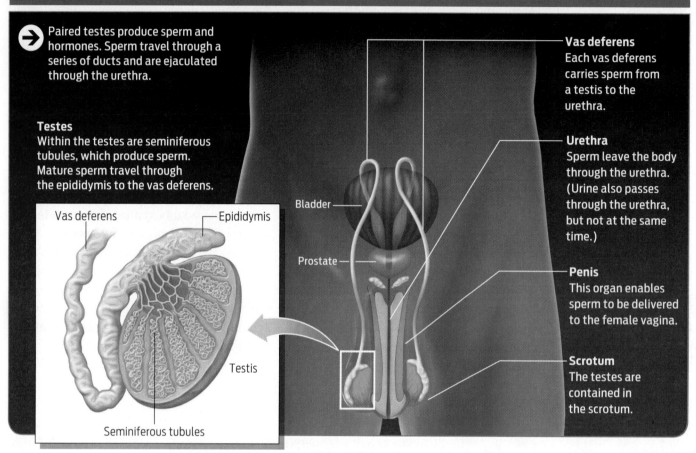

Paired testes produce sperm and hormones. Sperm travel through a series of ducts and are ejaculated through the urethra.

Testes
Within the testes are seminiferous tubules, which produce sperm. Mature sperm travel through the epididymis to the vas deferens.

Vas deferens

Epididymis

Testis

Seminiferous tubules

Bladder

Prostate

Vas deferens
Each vas deferens carries sperm from a testis to the urethra.

Urethra
Sperm leave the body through the urethra. (Urine also passes through the urethra, but not at the same time.)

Penis
This organ enables sperm to be delivered to the female vagina.

Scrotum
The testes are contained in the scrotum.

Conception and Infertility

During sex, when a man ejaculates into a woman's **vagina,** sperm swim up the reproductive tract, through the opening of the uterus, called the **cervix** (so called from the Latin word for "neck"), and into the oviducts. If a single sperm fuses with an egg, the egg is **fertilized** (Infographic 28.3). The fertilized egg, called a **zygote,** then travels into the uterus, where–if a successful pregnancy is to occur–it must implant into the uterine wall and eventually develop into a fetus.

That's the way normal conception happens–but several things can go wrong along the way. Physical damage to the reproductive organs can prevent fertilization. In some cases, a woman's oviducts can be blocked or damaged, preventing eggs from entering the

uterus or sperm from getting into the oviduct, where fertilization normally takes place. The most common cause of blocked tubes is pelvic inflammatory disease, which can be caused by sexually transmitted diseases such as chlamydia, a bacterial infection. An untreated infection can cause scar tissue to build up in the oviducts and block them.

In men, any obstructions in the vas deferens or epididymis will block sperm transport. Varicose veins in the testicles and sexually transmitted bacterial infections such as chlamydia or gonorrhea can also cause blocked tubes.

Fertility clinics can test for physical blockages and, in some cases, surgically correct them. When surgery isn't an option, IVF is often recommended. In IVF, hormones are

VAGINA
The first part of the female reproductive tract, extending up to the cervix; also known as the birth canal.

CERVIX
The opening or "neck" of the uterus, where sperm enter and babies exit.

FERTILIZATION
The fusion of an egg and a sperm to form a zygote.

ZYGOTE
A fertilized egg.

Fertilization Occurs in the Female Oviduct

During intercourse, sperm ejaculated from the penis enters the female reproductive tract. Sperm must swim through the tract into the oviducts to fertilize an ovulated egg. While many sperm may make it to the oviduct, only one will actually fertilize the egg. Blockages in the female reproductive tract can impede sperm passage to an egg, and consequently compromise fertilization.

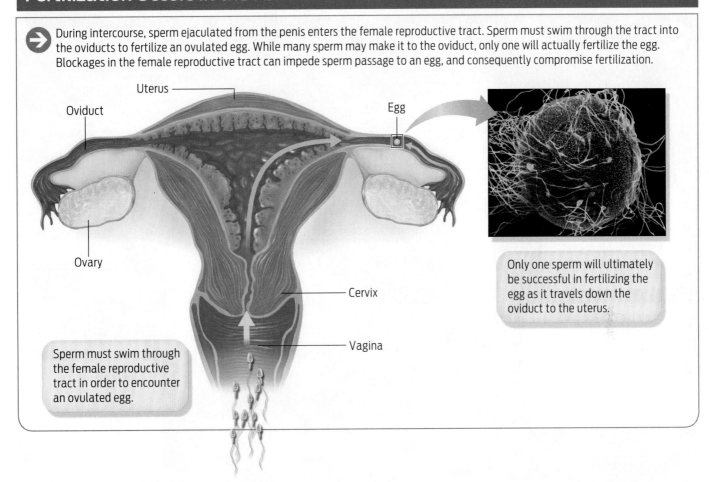

Uterus

Oviduct

Ovary

Egg

Cervix

Vagina

Sperm must swim through the female reproductive tract in order to encounter an ovulated egg.

Only one sperm will ultimately be successful in fertilizing the egg as it travels down the oviduct to the uterus.

EMBRYO
An early stage of development, an embryo forms when a zygote undergoes cell division.

administered to a woman to promote egg development, then eggs are extracted from her ovaries through a long, thin needle inserted through the vagina. Sperm are extracted from ejaculate or, in cases of physical blockage, from the epididymis. The sperm and eggs are combined outside the body in a petri dish to allow the sperm to fertilize the eggs. The resulting **embryos** are then inserted into the woman's uterus in hope that at least one will develop into a fetus **(Infographic 28.4)**. IVF has been used to help infertile couples conceive children since 1978, when Louise Brown, the first IVF baby, was born.

IVF may also be recommended in cases of low sperm number, abnormal sperm, and even in cases in which the cause of infertility can't be determined. In such cases the term "infertility" can actually be a misnomer. Many couples with defective sperm or unexplained infertility can still conceive a child naturally—it just may take longer. But because no one can predict how long it might take to achieve a successful pregnancy and because fertility decreases with age, IVF makes conception more likely by bringing sperm and egg together artificially.

Hormones and Fertility

Physical blockages and sperm quality account for only about 30% to 40% of all cases of infertility; another 20% to 30% are caused by hormonal

In Vitro Fertilization (IVF)

→ In vitro fertilization involves extracting eggs and sperm and mixing them together outside the body to allow fertilization. The resulting embryos are then inserted back into a woman's uterus in hope that at least one of them will implant into the uterus and grow into a fetus.

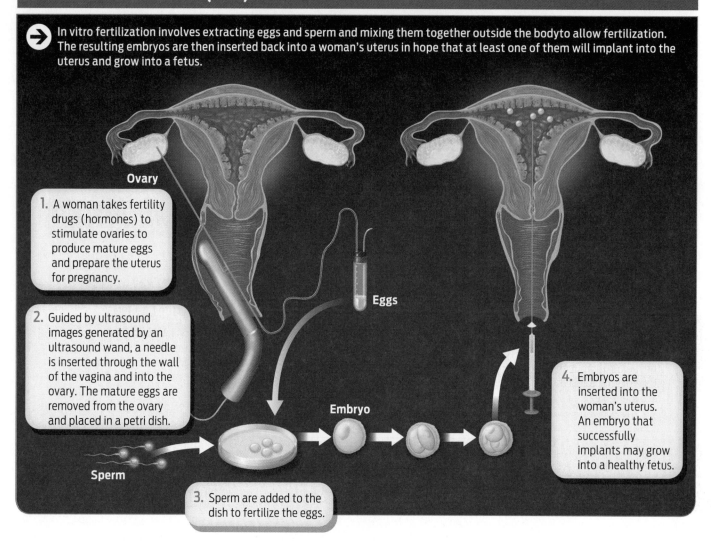

Ovary

1. A woman takes fertility drugs (hormones) to stimulate ovaries to produce mature eggs and prepare the uterus for pregnancy.

2. Guided by ultrasound images generated by an ultrasound wand, a needle is inserted through the wall of the vagina and into the ovary. The mature eggs are removed from the ovary and placed in a petri dish.

Eggs

Embryo

4. Embryos are inserted into the woman's uterus. An embryo that successfully implants may grow into a healthy fetus.

Sperm

3. Sperm are added to the dish to fertilize the eggs.

imbalances. Hormones are responsible for regulating the production of gametes, both sperm and egg. In females, estrogen and progesterone are key reproductive hormones without which a woman would be infertile. In males, testosterone is the primary hormone that stimulates sperm to develop.

In females, estrogen and progesterone are responsible for the menstrual cycle, a reproductive cycle that repeats roughly once every 28 days after the onset of puberty. During each cycle, estrogen and progesterone levels rise and fall to trigger the ovaries to release an egg and

prepare a woman's uterus for pregnancy should an egg be fertilized.

The hypothalamus controls levels of estrogen and progesterone in the body. The hypothalamus is the master coordinator of a number of important physiological functions, including hunger, thirst, and temperature regulation—in addition to reproduction. To regulate fertility, the hypothalamus secretes hormones that signal the **anterior pituitary gland** to produce two other key hormones. These two hormones, follicle-stimulating hormone and luteinizing hormone, travel through

ANTERIOR PITUITARY GLAND
The gland in the brain that secretes luteinizing hormone (LH) and follicle-stimulating hormone (FSH).

FOLLICLE-STIMULATING HORMONE (FSH)
A hormone secreted by the anterior pituitary gland. In females, FSH triggers eggs to mature at the start of each monthly cycle.

FOLLICLE
The part of the ovary where eggs mature.

LUTEINIZING HORMONE (LH)
A hormone secreted by the anterior pituitary gland. In females, a surge of LH triggers ovulation.

OVULATION
The release of an egg from an ovary into the oviduct.

CORPUS LUTEUM
The structure in the ovary that remains after ovulation and secretes progesterone.

MENSTRUATION
The shedding of the uterine lining (the endometrium) that occurs when an embryo does not implant.

HUMAN CHORIONIC GONADOTROPIN (hCG)
A hormone produced by an early embryo that helps maintain the corpus luteum until the placenta develops.

PLACENTA
A structure made of fetal and maternal tissues that helps sustain and support the embryo and fetus.

the bloodstream and directly stimulate the ovaries.

Follicle-stimulating hormone (FSH) acts on structures in the ovaries called **follicles,** each of which contains an immature egg. FSH signals follicles in the ovary to enlarge and to produce estrogen. Estrogen has several effects, one of which is to cause the endometrium to start to thicken. Estrogen also stimulates eggs within the ovaries to mature.

As hormone-secreting structures, the hypothalamus, the anterior pituitary, and the ovaries are all part of the endocrine system. All endocrine glands secrete hormones into the circulation. Hormones then travel though the bloodstream to reach their target cells. Hormones interact with their target cells by binding to specific receptors present on the target cells. For example, FSH binds to specific receptors on certain ovarian cells, triggering them to enlarge and release estrogen.

At about the midpoint of the cycle–roughly 10 to 14 days after a woman begins menstrual bleeding–rising estrogen levels trigger the brain to release a large amount of **luteinizing hormone (LH).** This LH surge triggers **ovulation**–the release of an egg from the follicle and the ovary itself into the oviduct. After the egg has been ovulated, the remaining follicle becomes a structure called the **corpus luteum,** which secretes progesterone. One of the most important roles of progesterone is to encourage the endometrium to continue to thicken. The thickened endometrium contains blood vessels and nutrients and is prepared to receive an embryo if the egg is fertilized.

Although both ovaries can release eggs during the same cycle, they typically take turns and only one egg is released during each cycle (identical twins occur when a single egg is released and fertilized and then splits into two embryos early in embryonic development). In about 1% of cycles, however, there are multiple ovulations, in which case fraternal twins, triplets–or as many embryos as there are ovulations–can develop.

Ovulation presents a crucial time window during which a woman can become pregnant. Sperm must swim through the cervix and uterus and into the oviduct containing the released egg in order to fertilize it. Because sperm can survive in the female reproductive tract anywhere from 3 to 7 days, a woman can become pregnant even if she has sex before she ovulates. Sperm can in effect "wait" in the oviduct for ovulation to occur. Once the egg leaves the oviduct, however, the odds that it will be fertilized are extremely small.

If an egg is not fertilized within 24 hours of being ovulated, it is no longer viable. The corpus luteum degenerates at about day 26 of the cycle, progesterone levels drop, and the uterine lining sloughs off, leading to **menstruation (Infographic 28.5).**

If the egg is fertilized it is called a zygote. As the zygote travels to the uterus it begins dividing and developing into an embryo. The embryo implants itself in the endometrium of the uterus about 1 week after the egg is fertilized. Once implanted in the uterus, the embryo secretes a hormone called **human chorionic gonadotropin (hCG),** which signals the corpus luteum to continue producing progesterone to support the thickening endometrium. This is the hormone that, in effect, tells the reproductive system that pregnancy has begun (hCG is the hormone that most pregnancy tests use as a marker to detect a pregnancy). Once the embryo implants, embryonic and maternal endometrial tissues interact to form the **placenta,** a disc-shaped structure that provides nourishment and support (like waste removal) to the developing fetus. In addition to its role in delivering oxygen, nutrients, and other key molecules (like antibodies from the mother that help protect the embryo against infections), the placenta eventually takes over the task of producing estrogen and progesterone from the corpus luteum. The ovaries stop responding to follicle-stimulating hormone and luteinizing hormone at the time of menopause, which in American women occurs at about age 51; at this time women stop

Hormones Regulate the Menstrual Cycle

A complex interplay of hormones from the hypothalamus, anterior pituitary gland, and ovaries drives the monthly female reproductive cycle.

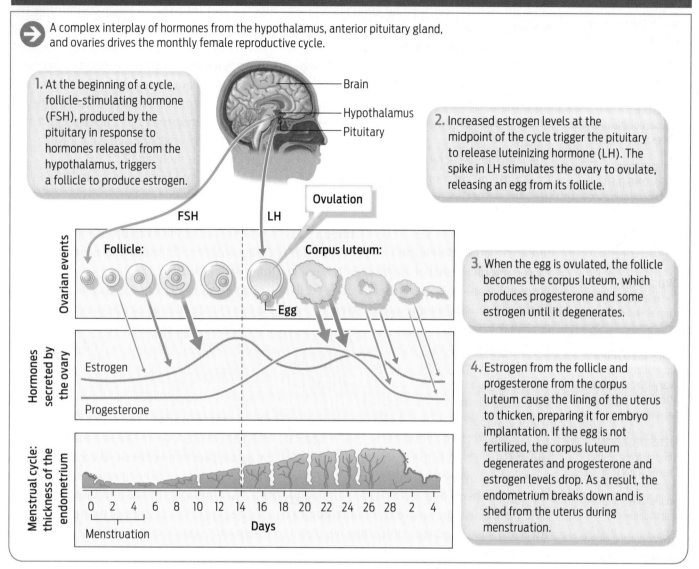

1. At the beginning of a cycle, follicle-stimulating hormone (FSH), produced by the pituitary in response to hormones released from the hypothalamus, triggers a follicle to produce estrogen.

2. Increased estrogen levels at the midpoint of the cycle trigger the pituitary to release luteinizing hormone (LH). The spike in LH stimulates the ovary to ovulate, releasing an egg from its follicle.

3. When the egg is ovulated, the follicle becomes the corpus luteum, which produces progesterone and some estrogen until it degenerates.

4. Estrogen from the follicle and progesterone from the corpus luteum cause the lining of the uterus to thicken, preparing it for embryo implantation. If the egg is not fertilized, the corpus luteum degenerates and progesterone and estrogen levels drop. As a result, the endometrium breaks down and is shed from the uterus during menstruation.

Brain
Hypothalamus
Pituitary
Ovulation
FSH
LH
Ovarian events
Follicle:
Corpus luteum:
Egg
Hormones secreted by the ovary
Estrogen
Progesterone
Menstrual cycle: thickness of the endometrium
0 2 4 6 8 10 12 14 16 18 20 22 24 26 28 2 4
Days
Menstruation

ovulating and stop having monthly reproductive cycles, and are therefore no longer able to conceive (Infographic 28.6).

Since hormones play such a crucial role in pregnancy and reproduction, many types of **contraception** make use of this phenomenon and interfere with the normal female hormone cycle (Table 28.1). In the combination birth control pill that contains both estrogen and progesterone, for example, these hormones are at levels that prevent the anterior pituitary gland from releasing follicle-stimulating and luteinizing hormones. This prevents ovulation

and consequently a woman taking this pill does not release eggs to be fertilized. Progesterone in birth control pills prevents successful pregnancy in other ways, too: it thickens the cervical mucus, blocking sperm from entering the uterus and oviducts, and also reduces endometrial thickening—a process that is necessary to support an embryo.

Men do not have a monthly hormone cycle, but beginning at puberty, sperm go through recognizable stages of development. The seminiferous tubules house precursor sperm cells that go through cell division—that is,

CONTRACEPTION
The prevention of pregnancy through physical, surgical, or hormonal methods.

Hormones Support Pregnancy

→ Estrogen and progesterone support the implanted embryo as it develops. Early in pregnancy, the embryo secretes human chorionic gonadatropin (hCG), which signals the corpus luteum to continue to produce these hormones. The placenta, once it has formed, takes over estrogen and progesterone production.

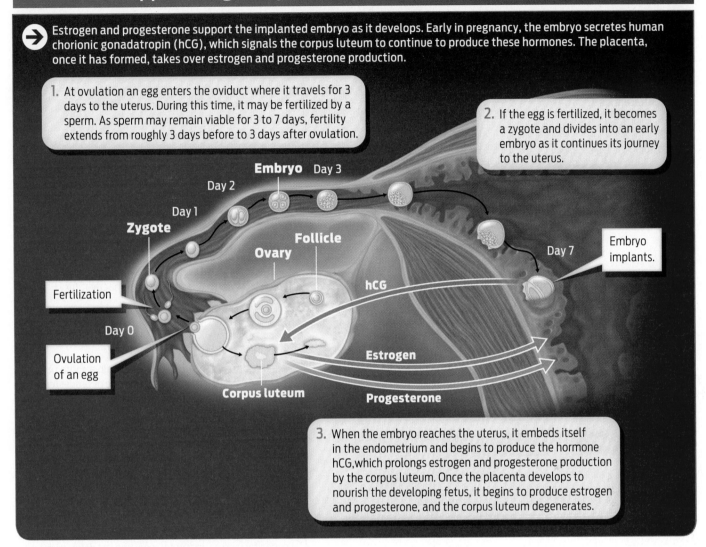

1. At ovulation an egg enters the oviduct where it travels for 3 days to the uterus. During this time, it may be fertilized by a sperm. As sperm may remain viable for 3 to 7 days, fertility extends from roughly 3 days before to 3 days after ovulation.

2. If the egg is fertilized, it becomes a zygote and divides into an early embryo as it continues its journey to the uterus.

3. When the embryo reaches the uterus, it embeds itself in the endometrium and begins to produce the hormone hCG, which prolongs estrogen and progesterone production by the corpus luteum. Once the placenta develops to nourish the developing fetus, it begins to produce estrogen and progesterone, and the corpus luteum degenerates.

meiosis—and specialization to produce mature sperm cells. It takes approximately 6 weeks for sperm to mature. Maturation is stimulated by testosterone, and although men produce testosterone continuously throughout their adult lives, they produce less of the hormone as they age and consequently sperm quality declines over time (Infographic 28.7).

The Many Causes of Infertility

Given how important hormones are to egg maturation and ovulation, it's not surprising that hormonal imbalances are a common cause of female infertility. Disruption in the part of the brain that regulates ovulation can cause low levels of luteinizing hormone and follicle-stimulating hormone, which in turn may prevent ovulation or make it erratic. Even slight irregularities in the hormone system can prevent the ovaries from releasing eggs. Specific causes of such hormonal imbalances include injury, tumors, excessive exercise, and starvation. Some medications are associated with ovulation disorders, and some studies have shown that stress can negatively affect fertility, as can poor nutrition.

TABLE 28.1

Contraction

 There are many ways to prevent conception. Currently available contraceptives include behavioral methods, physical and chemical barriers, hormones, and surgery.

METHOD	FAILURE RATE (% OF PREGNANCIES/ YEAR WITH EACH METHOD)	DESCRIPTION
No Contraception	85%	Sexual intercourse without any method of contraception.
Behaviors	3–27% 0% (abstinence)	The rhythm method (avoiding intercourse around the time a woman ovulates); withdrawal (the male withdraws his penis before ejaculating); sexual abstinence.
Barriers	2–21%	The male and female condom, the diaphragm, and the cervical cap. All of these prevent sperm from entering the uterus and are typically used with spermicidal foams or jellies that contain sperm-inactivating chemicals.
Hormones	0.3–8%	Female hormonal contraceptives contain a combination of synthetic estrogen and progesterone or progesterone only. Women can take these hormones in the form of combination pills, a skin patch, a cervical ring, a minipill, a regular injection, or an implant. All hormonal methods prevent pregnancy in three major ways: they thicken the cervical mucus, making it less likely that sperm will be able to swim through it and get into the oviducts; they prevent ovulation; and they thin the endometrium, so it cannot support the implantation of an embryo.
Surgery	0.5–0.15%	Surgical options include a vasectomy for men and tubal ligation for women. In a vasectomy, the vas deferens is cut, so sperm can no longer be ejaculated. In tubal ligation, the oviducts are cut and tied off, so sperm can no longer reach an ovulated egg. Both surgeries are essentially permanent, although they can be reversed with limited success. Those who do not want to have children, or any more children, typically choose surgical sterilization.
Intrauterine device	0.2–0.8%	An intrauterine device (IUD) is a small T-shaped device that typically contains copper or another metal. It is inserted into the base of the uterus through the cervix. The IUD is a long-term contraceptive option. It prevents pregnancy by thickening cervical mucus and consequently impeding sperm from entering the uterus. It also weakens the endometrium, making it less able to support an embryo.

Source: Trussell J. Contraceptive efficacy. In Hatcher RA, Trussell J, Nelson AL, Cates W, Stewart FH, Kowal D. *Contraceptive Technology: Nineteenth Revised Edition*. New York NY: Ardent Media, 2007.

How Sperm Form

→ Within seminiferous tubules in the testes, precursor cells are stimulated by testosterone to go through cell division (meiosis) and cell differentiation to develop into sperm. While the entire process takes approximately 6 weeks, because cells are in various stages of development a continuous supply of sperm is produced.

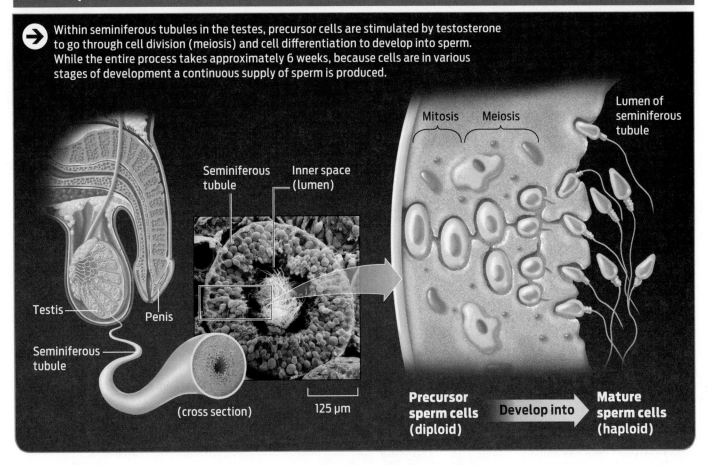

Seminiferous tubule

Inner space (lumen)

Mitosis   Meiosis

Lumen of seminiferous tubule

Testis

Penis

Seminiferous tubule

(cross section)

125 µm

Precursor sperm cells (diploid) → **Develop into** → **Mature sperm cells (haploid)**

Some women suffer from polycystic ovary syndrome, a condition associated with the production of excessive amounts of androgens (the "male" sex hormones, typically made in only small amounts in females). This syndrome is one of the most common hormonal disorders, affecting an estimated 10% of women of reproductive age. In addition to affecting ovulation, the condition is associated with ovarian cysts, irregular menstrual cycles, diabetes, and obesity.

Although men, too, can suffer hormonal imbalances, these are much less common than in women. Males may have low testosterone levels if they have a pituitary tumor that causes a reduction in the levels of FSH and LH (which in men are required to stimulate sperm and testosterone production by the testes), or some kind of damage to the testes. Injury, che-

motherapy, and radiation can also interfere with testosterone production. Low testosterone levels reduce both sperm production and libido.

Overall, there is no single factor that is the most common cause of infertility, but rather a combination of physical and hormonal abnormalities that together contribute to the 10% to 15% infertility rate we see in the U.S. population (**Infographic 28.8**). And there remains a small percentage of cases, approximately 10% of all infertile couples, in which the cause of infertility remains unknown.

The Problem of Multiples

Women with polycystic ovary syndrome or other hormonal disorders who wish to become pregnant are typically prescribed a course of

Causes of Infertility

In women:

Nonfunctional ovaries
Ovaries may fail to ovulate because of a variety of causes, including hormonal imbalances, genetic abnormalities, undeveloped ovarian tissue, endometriosis, and cancer.

Cysts, fibroids, polyps
Each of these is a type of abnormal growth that may block passages or interfere with normal function of the tissue.

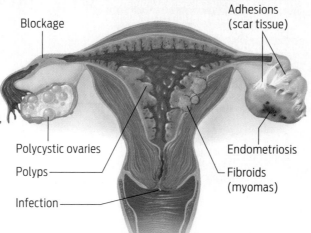

Blockage

Adhesions (scar tissue)

Polycystic ovaries

Polyps

Infection

Endometriosis

Fibroids (myomas)

Endometriosis
The tissue that lines the uterus grows abnormally and invades other tissues in the pelvis. The wayward tissue irritates the nerve endings of these organs and interferes with their function.

Nonreceptive fluids
Cervical mucus may be nonreceptive to sperm. Antisperm chemicals may be secreted.

Blockages
Passages may become blocked or disabled in both male and female reproductive systems because of tissue scarring, infection, cancer, or abnormal tissue growth.

Prostatitis
Sperm pass through and receive fluid from the prostate on their way into the urethra. An enlarged prostate can block the passage of semen.

In men:

Erectile dysfunction
Genetic abnormalities, neurological problems, hormonal imbalances, and physical blockages inhibit the ability of blood to flood the penis tissue to support an erection.

Testicular varicose veins
Valves in the veins that keep blood flowing in one direction deteriorate, causing blood to back up and pool. These enlarged veins can interfere with sperm production and transport.

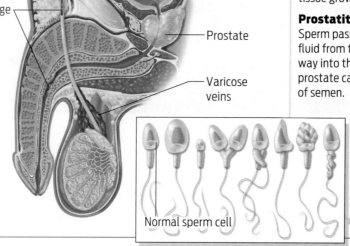

Blockage

Prostate

Varicose veins

Normal sperm cell

Sperm abnormalities
Men may have low numbers of healthy sperm and/or physically abnormal sperm.

---

fertility drugs. These are usually anterior pituitary hormones to stimulate multiple ovarian follicles to develop and multiple eggs to ovulate. Then, in a form of artificial insemination called **interuterine insemination,** sperm are injected into the uterus. Ultrasound technology is used to monitor ovulation, so the insemination can be performed when chances are highest that one or more eggs have been released into the oviduct.

Fertility drugs almost always cause two or more ovulations. This is desirable during IVF,

in which the number of eggs that are fertilized outside the body and implanted into the uterus can be controlled. But this advantage turns into a liability with intrauterine insemination. Since sperm are inserted into the uterus during insemination, it's difficult to control the number of eggs that are fertilized and, consequently, the number of conceptions. Multiple births result when more than one egg is fertilized (each by a different sperm), leading to more than one embryo implanting in the uterus and developing into a fetus **(Infographic 28.9).**

INTRAUTERINE INSEMINATION (IUI)
A form of assisted reproduction in which sperm are injected directly into a woman's uterus.

Assisted Reproductive Technologies Can Result in Multiple Births

One hazard of assisted reproduction is a high probability of multiple births. Babies born as multiples are more likely to be born underweight and premature, putting them at risk for a variety of serious health conditions.

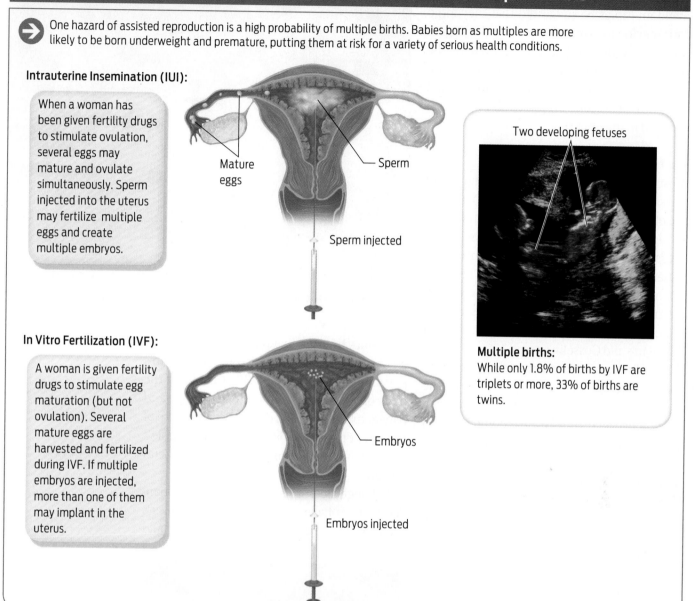

Intrauterine Insemination (IUI):

When a woman has been given fertility drugs to stimulate ovulation, several eggs may mature and ovulate simultaneously. Sperm injected into the uterus may fertilize multiple eggs and create multiple embryos.

Mature eggs

Sperm

Sperm injected

In Vitro Fertilization (IVF):

A woman is given fertility drugs to stimulate egg maturation (but not ovulation). Several mature eggs are harvested and fertilized during IVF. If multiple embryos are injected, more than one of them may implant in the uterus.

Embryos

Embryos injected

Two developing fetuses

Multiple births:
While only 1.8% of births by IVF are triplets or more, 33% of births are twins.

In Suleman's case IVF treatment was too effective. Fertility treatment was also too effective for Jon and Kate Gosselin, the famous couple who starred in the television reality series *Jon & Kate Plus 8.* The couple first had twins and then later sextuplets, all of whom were conceived through intrauterine insemination.

The book *Multiple Blessings,* by the Gosselins and Beth Carson, tells the story. Kate's first treatment consisted of "painful injections" of fertility drugs to stimulate her ovaries. She ended up having twins. A year later, Kate longed for another baby. This time, Kate saw a different specialist. Kate and her husband told their doctor that they did not want multiples but would not selectively "reduce," or abort a fetus. She received another round of fertility drugs.

When Kate had an ultrasound examination during the treatment, the Gosselins saw three, possibly four, developing eggs. Their doctor told them that they were unlikely to end up with three or four babies, but to be sure to avoid multiples he suggested that they forgo intrauterine insemination then and try again later. They decided to go ahead with the insemination anyway. When the sonogram showed seven developing embryos (one would later disappear), Jon dropped to his knees, Kate recalled in the book.

Costly Care

Most insurance policies don't cover the cost of assisted reproduction, and many patients pay out of pocket. Because the costs of treatment are so high, many doctors find themselves under pressure to be aggressive with treatment with patients who have limited resources. Therefore, financial considerations can dictate a course of treatment, even when the treatment is unlikely to be successful or to be too successful, as in Suleman's and Gosselin's cases.

As of 2010, a single insemination treatment cost an average of $6,000–almost half the cost of an average in vitro fertilization cycle. Since most health insurance companies do not cover assisted reproduction, couples are often left to decide on a procedure based on how much they can afford. Many couples may choose insemination over in vitro fertilization simply because it is less expensive, even though the procedure has a low success rate. And when it is successful, it carries a high risk of multiples.

The average cost of a single treatment is more than $10,000. Since each round of IVF carries only about a 40% success rate according to ASRM, some couples must undergo multiple rounds to achieve a successful pregnancy.

The success rate of a single round of intrauterine insemination is much lower. On average, only 10% to 20% of all insemination treatments result in a live birth. But younger women and women who take fertility drugs and, as a result, have multiple ovulations, tend to have higher success rates. This last point is key: multiple ovulations double the odds of a successful

Kate Gosselin and her eight children.

pregnancy by insemination. Similarly, the more embryos that are transferred into a woman's uterus after each round of IVF, the higher the odds of a pregnancy. In the end, doctor and patient together should decide which course of treatment to pursue on the basis of a couple's specific fertility condition and financial resources.

Although most of Suleman's health and financial records have remained confidential, it's unlikely that the unemployed 33-year-old could

Suleman (right) with some of her 14 children.

have afforded to pay for multiple rounds of IVF. The Associated Press and various newspapers have reported that Suleman received at least $165,000 in disability benefits from the state of California. But considering she already had six children to care for and had been unemployed for many years, it's unclear how her second round of IVF (which resulted in eight babies) was paid for.

Medical details revealed during a court case brought by the California Medical Board against Suleman's doctor, Michael Kamrava, show that Kamrava created 14 embryos and implanted a dozen of them. Only eight survived, and the babies were born 9 weeks prematurely. The Associated Press reported that during the hearing Kamrava said he regretted implanting the 12 embryos and "would never do it again."

Kamrava further stated that Suleman was adamant about using all 12 embryos, even though he suggested implanting only 4. "She just wouldn't accept doing anything else with those embryos. She did not want them frozen, she did not want them transferred to another patient in the future," he said, according to the AP story. Kamrava said that he consented only after Suleman agreed to have a fetal abortion if necessary. After the implantation, however, he only heard from Suleman after the birth of her octuplets, despite his numerous attempts to contact her.

From a medical perspective, the birth of multiples from any assisted reproduction procedure is problematic. According to the March of Dimes, 50% of twins and 90% of triplets are born prematurely, as are virtually all quads and quintuplets. Their lungs are often immature and so the babies must be hooked up to mechanical breathing ventilators, which sometimes scar the lungs so that these children will for the rest of their lives be prone to asthma, pneumonia, chronic lung disease, and other respiratory disorders. And because their brains aren't fully developed, premature babies are susceptible to brain hemorrhages and to developmental difficulties, including learning disabilities (see **Up Close: Prenatal Development**). These medical conditions pose a huge burden on the health care system.

To reduce the birth of multiples and prevent the associated health problems, the American Society for Reproductive Medicine recommends transferring no more than two embryos in women under 35 years of age (and only one if possible) and no more than five in women over 40, because as a woman ages the success rate tends to drop (Suleman had six times the number of recommended embryos implanted). These guidelines appear to have been effective: in 2007, only 1.8% of live births to patients under 35 were triplets or more, as compared to 6.4% in 2003. But assisted reproduction still results in a high number of twins—in 2007, 33% percent of births from IVF alone were twins.

Despite the drop in multiple births, some lawmakers want more control. The risk that doctors may act against the ASRM guidelines, with negative consequences, remains. The California Medical Board continues to seek to revoke Kamrava's medical license, claiming that he was negligent not only in Suleman's case, but also in the cases of two other women who suffered serious medical complications because of aggressive fertility treatments. Lawmakers are concerned that there is nothing to prevent cases like these from happening again. Several states are considering legislation that would regulate doctors. A Missouri bill, for example, would require doctors not to exceed the ASRM's embryo-transfer guidelines. But experts fear that regulation would hinder good care. Fertility doctors need flexibility to tailor treatment to a couple's individual condition of fertility, they argue. Such legislation "seeks to substitute the judgment of politicians for that of physicians and their patients," ASRM president R. Dale McClure has said in a statement evaluating a similar bill.

Instead of putting controls on doctors, some groups favor regulation that would broaden access to treatment—legislation that would require health insurers to cover infertility diagnostics and treatment, for example. As long as fertility treatment continues to be a financial burden on couples, fertility doctors will face pressure from patients to give them the most for their money, says Barbara Collura, executive director of RESOLVE, an infertility advocacy

The embryonic stage begins 1 week after the embryo implants and continues until the 8th week after fertilization. After that, the developing baby is known as a fetus. Pregnancy is divided into three trimesters.

1st Trimester: development of tissue layers and vital organs

Embryonic stage

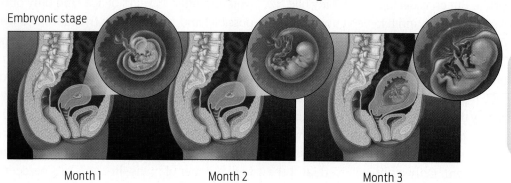

Month 1   Month 2   Month 3

The first trimester includes the embryonic stage of development and the early fetal stage. The embryo and fetus grow rapidly and critical organs develop.

2nd Trimester: growth and gender determination

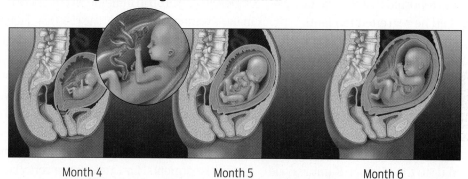

Month 4   Month 5   Month 6

During the second trimester, the fetus continues to grow and develop features, including external genitalia.

3rd Trimester: weight gain and organ system development

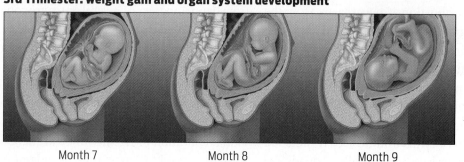

Month 7   Month 8   Month 9

During the third trimester the fetus becomes fully developed and gains sufficient weight to be born at full term about 35 weeks after fertilization.

group. Broad health insurance coverage would eliminate the cost factor. The result might be that couples would be able to forgo treatments such as insemination that have high risks with low success rates and skip directly to IVF when appropriate. Cost would not be a factor when considering treatment options.

"If nothing else, these high profile cases have served as a

> "If nothing else, these high profile cases have served as a wake-up call."
> – Barbara Collura

wake-up call," says Collura. The community of health care workers is examining its own procedures and methods because it would rather self-regulate than have regulation imposed on it from the outside, she adds. "After Suleman, the community is really looking into how it [the birth of octuplets] happened and how it can prevent it from happening again." ■

How Do Other Organisms Reproduce?

All organisms reproduce, but not all organisms have sex. Reproduction is a fundamental facet of all life–without the ability to reproduce any species would die out. Humans reproduce sexually, which means that reproduction involves two parents: one provides eggs, and the other provides sperm. Not all organisms reproduce in the same way, nor do they have the same sexual anatomy as humans.

In humans and other mammals, fertilization of an egg by sperm occurs internally, in a female oviduct. This system allows sperm and egg to be protected but limits the number of eggs that can be fertilized at any one time and requires a "sperm delivery system" (that is, a penis) to place the sperm in a location where they can easily swim to the egg.

Many nonmammalian species also rely on internal fertilization, but via a different sexual anatomy. Internal fertilization in birds and reptiles, for example, occurs in an organ called the cloaca (plural: cloacae). The cloaca is the shared opening for solid waste, urine, and the reproductive system. Birds briefly touch their cloacae together (in the so-called cloacal kiss) and this speedy kiss-and-run is sufficient for sperm to be transferred from the male to the female.

Birds Reproduce Sexually and Fertilize Eggs Internally

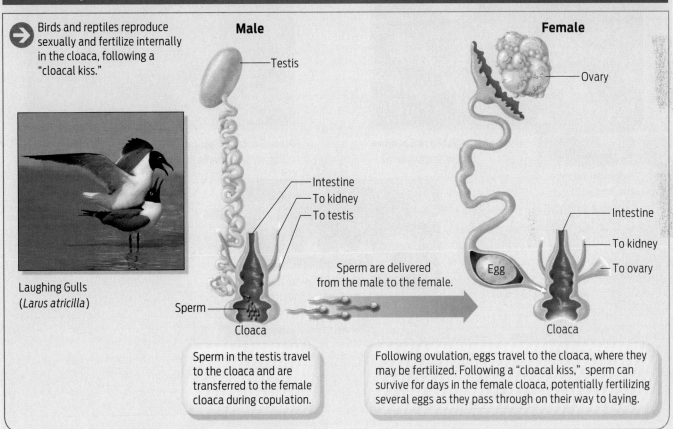

→ Birds and reptiles reproduce sexually and fertilize internally in the cloaca, following a "cloacal kiss."

Laughing Gulls
(*Larus atricilla*)

Male

Testis
Intestine
To kidney
To testis
Sperm
Cloaca

Sperm are delivered from the male to the female.

Female

Ovary
Intestine
To kidney
To ovary
Egg
Cloaca

Sperm in the testis travel to the cloaca and are transferred to the female cloaca during copulation.

Following ovulation, eggs travel to the cloaca, where they may be fertilized. Following a "cloacal kiss," sperm can survive for days in the female cloaca, potentially fertilizing several eggs as they pass through on their way to laying.

Interestingly, as embryos human females have a cloaca, but the single cloacal opening divides into separate openings during embryonic development. In very rare cases, baby girls are born with an intact cloaca–having only one opening, rather than the usual three. Surgical correction to form distinct anal, vaginal, and urethral openings is possible in these cases.

In contrast, many organisms rely on external fertilization– their gametes fuse in the outside environment. Female salmon, for example, deposit their eggs in gravel nests in streambeds. Male salmon then swim over the eggs and release sperm, fertilizing thousands of eggs simultaneously. Other aquatic species such as coral and hydra also reproduce this way. Many plant species also practice external fertilization.

Fish Reproduce Sexually and Fertilize Eggs Externally

Organisms like fish reproduce sexually, with two parents each contributing a gamete. The gametes combine externally during fertilization.

Female Sockeye Salmon
(*Oncorhynchus nerka*)

Male Sockeye Salmon
(*Oncorhynchus nerka*)

Eggs are laid by the female in the river rock bed.

Sperm are deposited by the male over the eggs.

Eggs and sperm mix in the river gravel and dozens of eggs are fertilized externally.

Still other species use an entirely different form of reproduction altogether: they reproduce asexually. In asexual reproduction, a single parent produces offspring without additional genetic input from another individual. All bacteria reproduce asexually. Some organisms, such as certain types of fungi, can reproduce both sexually and asexually. Baker's yeast (*Saccharomyces cerevisiae*) is a fungus that is commonly used to make bread rise. *S. cerevisiae* can produce gametes called spores, and these gametes can fuse to generate zygotes that develop into unicellular yeast. But both the spores and the yeast can also make exact copies of

themselves by mitotic cell division, resulting in identical populations of cells.

Why is there so much variation in the way that organisms reproduce? The short answer, as always, is evolution. Natural selection has favored adaptations that allow organisms to thrive in their particular environments. For example, while asexual reproduction produces offspring that are genetically identical to one another and to the parent, this type of reproduction is generally rapid and allows populations to grow quickly, which is an advantage if you are a bacterium or have a short life span or are easily killed.

Bacteria Reproduce Asexually

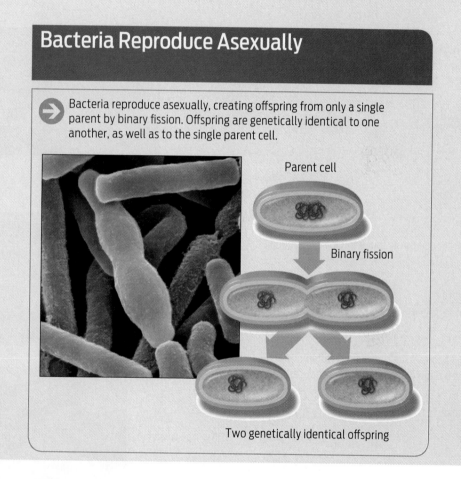

→ Bacteria reproduce asexually, creating offspring from only a single parent by binary fission. Offspring are genetically identical to one another, as well as to the single parent cell.

Parent cell

Binary fission

Two genetically identical offspring

Relative to external fertilization, internal fertilization places a large demand on the mother because she not only supplies all nutrients to a growing fetus, but also provides a protective environment within her body during the long time it takes for a fetus to grow into a baby. However, this maternal investment, while leaving the mother vulnerable during the pregnancy, often results in the successful live birth of a baby from each egg that is fertilized, compared to the exposed embryos that result from external fertilization. Over time, different organisms have adapted different reproductive strategies to ensure that their offspring survive in the particular environment each organism occupies.

Yeast Reproduce Both Sexually and Asexually

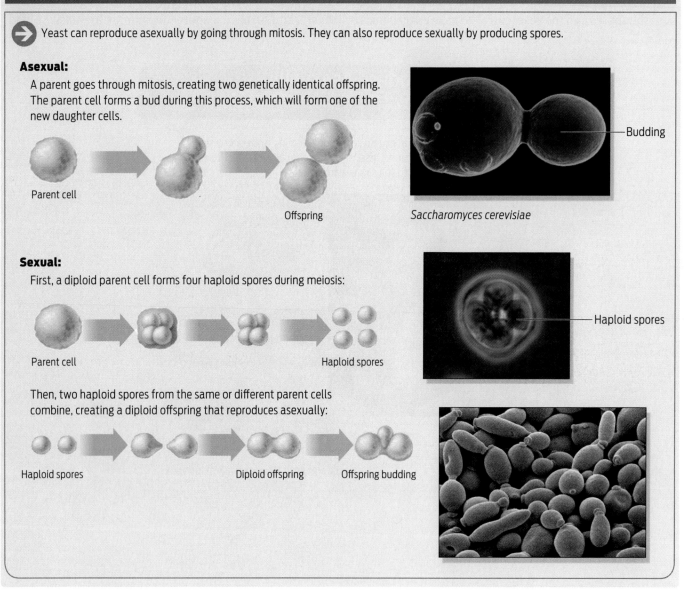

Yeast can reproduce asexually by going through mitosis. They can also reproduce sexually by producing spores.

Asexual:

A parent goes through mitosis, creating two genetically identical offspring. The parent cell forms a bud during this process, which will form one of the new daughter cells.

Parent cell

Offspring

Budding

Saccharomyces cerevisiae

Sexual:

First, a diploid parent cell forms four haploid spores during meiosis:

Parent cell

Haploid spores

Haploid spores

Then, two haploid spores from the same or different parent cells combine, creating a diploid offspring that reproduces asexually:

Haploid spores

Diploid offspring

Offspring budding

▶ Summary

■ The female reproductive system consists of paired ovaries, which produce eggs and secrete the hormones estrogen and progesterone, as well as accessory structures that enable fertilization and support pregnancy.

■ The male reproductive system consists of paired testes, which produce sperm, the hormone testosterone, as well as accessory structures that enable fertilization.

■ Fertilization of an egg by a sperm occurs in a female oviduct. Only one sperm can fertilize an egg at one time.

■ The female reproductive cycle is coordinated by a complex balance of hormones that is controlled by the brain.

■ Hormones from the hypothalamus trigger the anterior pituitary gland to release follicle-stimulating hormone and luteinizing hormone, which in turn stimulate eggs to develop and the ovary to secrete estrogen and progesterone.

■ Ovulation, the monthly release of an egg from an ovarian follicle, is caused by a spike in LH at the midpoint of the cycle.

■ Estrogen and progesterone stimulate eggs to develop and the endometrium to thicken and prepare for a possible pregnancy.

■ Upon fertilization, the zygote divides and travels to the uterus, where it implants in the nutrient-rich endometrium.

■ The implanted embryo secretes the hormone hCG, which acts on the corpus luteum to maintain progesterone secretion.

■ If an egg is not fertilized, progesterone levels fall and the endometrium sloughs off during menstruation.

■ Sperm develop in the seminiferous tubules of the testes. The process is stimulated by testosterone.

■ Infertility has many causes, including blocked passageways caused by infection and scar tissue, genetic abnormalities, and hormonal deficiencies.

■ Assisted reproduction involves artificially bringing sperm and egg together, either inside the body (in IUI) or outside the body (in IVF).

■ Fertility drugs increase the number of eggs that mature and are ovulated by a female at one time. Multiple pregnancies result when sperm fertilize more than one available egg.

■ Humans and other mammals, as well as birds and reptiles, reproduce sexually using internal fertilization. Other sexually reproducing animals, such as fish, use external fertilization. Some organisms are asexual and have no sex at all.

REPRODUCTIVE ANATOMY

Human males and females have distinct reproductive tracts that enable successful fertilization and pregnancy.

HINT See Infographics 28.1–28.3 and 28.7.

⊘ KNOW IT

1. Sperm develop in
   a. the epididymis.
   b. the vas deferens.
   c. the seminiferous tubules.
   d. the urethra.
   e. the penis.

2. Why can untreated pelvic inflammatory disease lead to infertility?
   a. because it prevents ovulation
   b. because it scars and blocks the oviducts
   c. because it scars and blocks the cervix
   d. because it interferes with estrogen production by the ovaries
   e. because it interferes with FSH and LH production by the anterior pituitary gland

3. Describe the relationship between (a) the uterus and the cervix and (b) the uterus and the endometrium.

⊘ USE IT

4. List the structures that sperm must pass through to reach and fertilize an egg. Begin with the seminiferous tubules.

5. A friend tells you that her boyfriend has been diagnosed with gonorrhea, a sexually transmitted disease. She isn't worried for herself because she doesn't have any symptoms of infection. What can you tell her about the invisible risks of an untreated sexually transmitted bacterial infection?

HORMONES AND REPRODUCTION

A complex interplay of hormones regulates gamete development and supports a successful pregnancy.

HINT See Infographics 28.5 and 28.6.

⊘ KNOW IT

6. What is the source—testes, ovaries, anterior pituitary gland, or embryo—of each of the following hormones?
   Luteinizing hormone (LH)
   Follicle-stimulating hormone (FSH)
   Testosterone
   Estrogen
   Progesterone
   hCG

7. The hormone hCG is an indicator of pregnancy; it also
   a. signals the corpus luteum to keep producing progesterone.
   b. triggers ovulation.
   c. acts on the anterior pituitary gland, causing it to release a surge of LH.
   d. acts on the endometrium, causing it to thicken.
   e. attracts sperm.

⊘ USE IT

8. Which of the following would most directly cause reduced levels of estrogen production?
   a. an anterior pituitary tumor that increases secretion of LH
   b. an increase in hypothalamus hormones that target the anterior pituitary
   c. an anterior pituitary tumor that increases secretion of FSH
   d. a decrease in hypothalamus hormones that target the anterior pituitary
   e. anterior pituitary damage that prevents synthesis and release of FSH

9. In an episode of *Law & Order: Special Victims Unit,* a blood sample from a crime scene was found to have extremely low levels of FSH and LH. Detectives used this information to determine that the blood came from a prepubescent girl, not a woman of reproductive age. Explain how they reached this conclusion.

10. As discussed in this chapter, oral contraceptives (such as the combination birth control pill) are designed to block ovulation in women. As males do not ovulate, a male hormonal contraceptive would

have to target sperm development. Which hormone would have to be blocked to prevent sperm development in males? What would be a likely undesired consequence of this type of male contraception?

INFERTILITY AND ASSISTED REPRODUCTION

There are many possible causes of infertility. Successful pregnancy can be achieved through a variety of assisted reproductive techniques, each of which has its own challenges, risks, and benefits.

HINT See Infographics 28.4, 28.8, and 28.9.

➔ KNOW IT

11. Which of the following could interfere with ovulation?

- **a.** blocked oviducts
- **b.** chronically low levels of LH
- **c.** excessive production of cervical mucus that blocks the cervix
- **d.** presence of sperm in the oviduct
- **e.** low levels of hCG

12. Compare and contrast in vitro fertilization (IVF) and intrauterine insemination (IUI).

➔ USE IT

13. Assume that there is an array of diagnostic methods available to you, including blood tests to determine hormone levels and ultrasound to visualize internal structures. What results might confirm each of the following infertility-associated conditions? Be as specific as possible.

- **a.** a blocked epididymis
- **b.** polycystic ovary syndrome
- **c.** menopause
- **d.** oviduct scarring

14. Why does IUI have a higher risk of multiple births than IVF?

SCIENCE AND SOCIETY

15. What do you think about Suleman's decision to have 14 children? Should local authorities or the federal government be empowered to regulate the number of children someone is allowed to have?

16. From the perspective of a fertility specialist, how would you respond to a congressional representative about a proposed increased regulation of fertility clinics? In order to make a convincing argument, include both pros and cons, medical and scientific considerations, and a description of the patient population that this specialist serves.

Some Important Features of Innate Immunity

a. Physical barriers

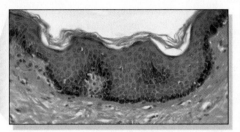

Skin (on the outside of the body) and mucous membranes (lining the inside of the body) have layers of tightly packed cells that prevent pathogens from entering the body. The mucus that coats mucous membranes traps foreign substances.

b. Inflammation

1. Microbes get past physical barriers.
2. Damaged cells and microbes release molecules that increase blood flow and attract white blood cells to infected areas.
3. Blood vessels leak, causing surrounding tissue to swell with fluid that contains clotting factors and white blood cells.
4. White blood cells ingest pathogens and also trigger an adaptive response. Clotting reactions contain the infection.

c. Phagocytes

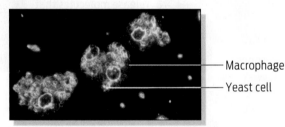

Macrophage
Yeast cell

Macrophage engulfs yeast cells

Phagocytes (including macrophages and neutrophils) can recognize, bind to, and ingest pathogens. Phagocytes also trigger inflammation and adaptive immune responses.

d. Antimicrobial chemicals

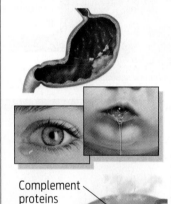

Acid in the stomach kills many of the microorganisms that we ingest.

Tears and **saliva** contain an enzyme that breaks down the cell walls of bacteria, causing them to burst.

Complement proteins

Complement proteins in blood puncture holes in bacterial cells or coat the cell's surface, making them more easily destroyed by phagocytes.

---

of this balance that played a crucial role in the 1918 pandemic. In the face of such an aggressive flu virus, the inflammatory response was so massive that it damaged the very tissue it was supposed to protect. As fluids leaked out of blood vessels, they impaired oxygen uptake in the lungs; and as phagocytes produced chemicals to kill pathogens, those same chemicals destroyed host cells and tissues and destroyed lung tissue. What killed

The inflammatory response was so massive that what killed many people was not the virus itself, but the overly aggressive inflammatory response.

many people was not the virus itself but the overly aggressive inflammatory response, according to research described by John Barry in his 2004 book, *The Great Influenza* (Infographic 29.5).

Others died not from the virus itself or its respiratory consequences but from secondary infections. Because influenza weakened the defenses of the respiratory tract, other organisms, such as bacteria, had unimpeded entry into the lungs. Bacterial pneumonia—an infection of the lungs—was responsible for at least half of all deaths during the 1918 pandemic, according to research by Jeffery Taubenberger, a virologist at the U.S. Centers for Disease Control and Prevention.

Lung Inflammation Can Kill

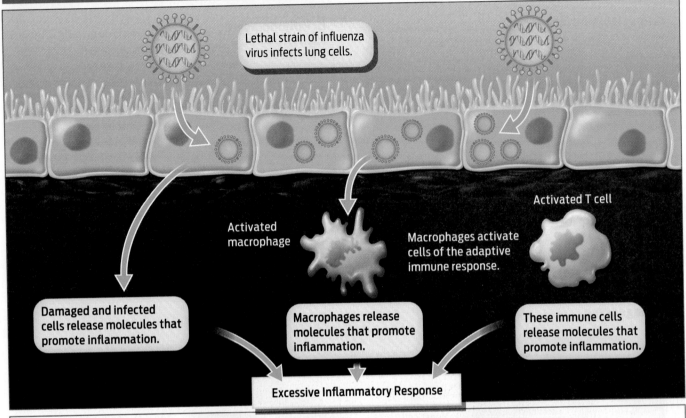

Lethal strain of influenza virus infects lung cells.

Activated macrophage

Macrophages activate cells of the adaptive immune response.

Activated T cell

Damaged and infected cells release molecules that promote inflammation.

Macrophages release molecules that promote inflammation.

These immune cells release molecules that promote inflammation.

Excessive Inflammatory Response

Acute respiratory distress:

- Cell death and debris
- Dilation of blood vessels
- Massive fluid influx
- Influx of white blood cells
- Loss of gas exchange
- Pneumonia

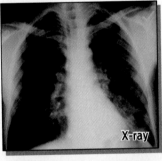

X-ray

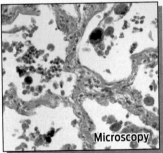

Microscopy

Infected lung is cloudy on x-ray and filled with immune cells, fluid, and debris, preventing normal function.

Normal lung:

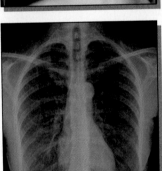

X-ray

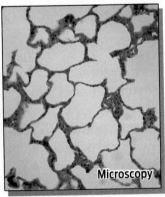

Microscopy

Normal lung is clear on x-ray and has space between cells that can fill with air.

Immunological Memory

Not everyone who became infected with the 1918 flu died, despite its virulence. In fact, while there were approximately 50 million deaths, experts estimate that some 525 million people were infected. How so many people were able to fight the infection while others died remains mostly a mystery. But scientists do have some clues.

Of those who became infected and then recovered, some may have had partial immunity from

B CELLS
White blood cells that mature in the bone marrow and produce antibodies during the adaptive immune response.

Memory Cells Mount an Aggressive Secondary Response

➡️ The adaptive immune system's primary humoral response is slow and produces low levels of antibodies. Upon subsequent exposures, memory cells produced during the primary response respond quickly and trigger B cells to produce high levels of antibodies. This secondary response is so rapid and effective that illness does not occur. The secondary response of cell-mediated immunity and cytotoxic T cells is also quick and effective.

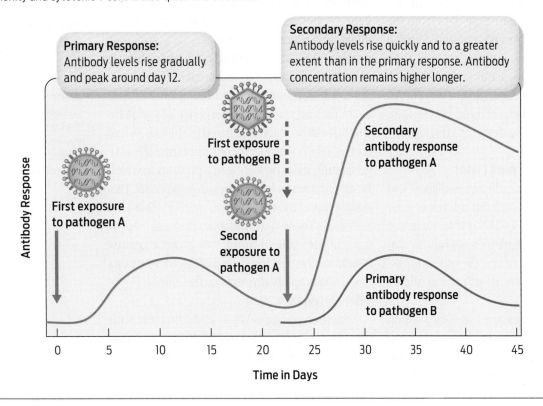

Primary Response:
Antibody levels rise gradually and peak around day 12.

Secondary Response:
Antibody levels rise quickly and to a greater extent than in the primary response. Antibody concentration remains higher longer.

First exposure to pathogen B

Secondary antibody response to pathogen A

First exposure to pathogen A

Second exposure to pathogen A

Primary antibody response to pathogen B

Antibody Response

Time in Days
0   5   10   15   20   25   30   35   40   45

PRIMARY RESPONSE
The adaptive response mounted the first time a particular antigen is encountered by the immune system.

MEMORY CELL
A long-lived B or T cell that is produced during the primary response and that is rapidly activated in the secondary response.

SECONDARY RESPONSE
The rapid and strong response mounted when a particular antigen is encountered by the immune system subsequent to the first encounter.

tissues. An autoimmune disease is the body's inappropriate immune response against itself.

Building a Line of Defense

How do the innate and adaptive immune systems work together? First-time exposure to a pathogen will almost certainly cause illness, because the adaptive response takes 7 to 10 days to develop. Over time an exposed individual will recover as T and B cells are activated and antibody levels increase. This initial slow response is the **primary response.** As B and T cells are churned out, some of them become **memory cells.** These memory cells remain in the bloodstream and "remember" the infection. The next time the same pathogen is encountered, memory B and T cells become active, dividing rapidly and producing very high levels of antibodies. They

fight the specific pathogen so quickly that the illness usually doesn't occur a second time. This rapid reaction is called the **secondary response** (Infographic 29.9).

The secondary response is what makes us immune to a particular infection. It's also how **vaccines** work. The source of all vaccines is the pathogen itself–some vaccines are made of only the antigens that cause an immune response, while others are a weakened or essentially dead version of the entire pathogen. However the vaccine is made, the goal is the same: to create a primary response in the body that's strong enough to create memory cells, yet weak enough not to cause disease symptoms. Thus, if the pathogen is subsequently encountered naturally, the secondary response is prepared. In this way, vaccination is like being infected with a pathogen without having the disease.

Even if they haven't been vaccinated, people exposed to a pathogen that is similar to a pathogen with which they were previously infected may be partly protected from the disease caused by the new pathogen. Memory B and T cells may still respond, although only partially–in which case the illness may occur, but mildly.

Evidence suggests that such partial immunity may have helped some of those infected survive the 1918 flu pandemic. Statistical data from the time show that people over 65 accounted for the fewest influenza cases, suggesting that they might over the years have acquired immunity or partial immunity from earlier infection. Partial immunity might have helped these people fight off the virus before it dug deep into the lungs.

Lessons Learned 75 Years Later

When Kawaoka and his colleagues discovered that the 1918 flu virus carried alleles of four genes that enabled the virus to replicate in the lower respiratory tract, thus making the disease so deadly, a piece of the 75-year-old mystery was solved. But where did these alleles come from?

New influenza viruses are constantly being produced by two mechanisms: by mutation and by grabbing genes from other viruses. Because influenza viruses replicate their genetic material so rapidly and don't "proofread" the replicated copies, mistakes often occur in the genomes of newly replicated viruses. **Antigenic drift** is the gradual accumulation of mutations that causes small changes in the antigens on the virus surface. Antigenic drift explains why there can be different types, or strains, of influenza circulating at the same time.

Two important antigens on the influenza virus are hemagglutinin (which binds to receptors on host cells and enables the virus to enter host cells) and neuraminidase (which helps new viruses exit host cells). The host immune system mounts an adaptive response specifically to these two antigens. When there is a change in hemagglutinin or neuraminidase or both, the memory cells no longer recognize them. These new antigens prompt a new and slow primary immune response. Anti-genic drift causes the seasonal variation in circulating flu viruses.

An influenza virus strain can also swap genes with other strains of influenza. While every influenza strain contains the same set of genes, the particular alleles of these genes that are present differ from strain to strain. The exchange of alleles between two strains that have infected the same cell does not simply create a small change in viral gene sequence: it introduces an entirely new allele, and therefore an entirely new antigenic protein. This process, called **antigenic shift,** is responsible for pandemic outbreaks of flu, including the 1918 pandemic, and the appearance of avian flu and the 2009 emergence of swine flu (H1N1), both of which originated in animals. When a new strain of influenza emerges through antigenic shift, it has new antigenic proteins to which humans have not been previously exposed. This means that they will not have memory cells that recognize these new antigens. Because people have no existing immunity to protect against infection by emerging strains of flu, these strains can spread rapidly throughout the human population (**Infographic 29.10**).

Together, antigenic drift and antigenic shift create an increasing variety of strains over time until one of the variants is able to infect human cells so efficiently that it sweeps through the population and causes a pandemic. These strains are named by the type of H (hemagglutinin) and N (neuraminidase) proteins they carry, with some antigenic combinations more deadly than others. Tracking the strains as they move through the population helps public health officials predict how severe a coming flu season will be. But since viruses can mutate every time they reproduce, they continue to evolve during epidemics. This unpredictability is what makes influenza so frightening: an apparently mild outbreak can suddenly become deadly.

> This unpredictability is what makes influenza so frightening: an apparently mild outbreak can suddenly become deadly.

Preventing Pandemics

Because influenza viruses change so quickly, public health officials closely monitor the types of influenza viruses that infect animals. Viruses that infect animals may accumulate mutations or pick up genes that enable them to infect human cells.

VACCINE
A preparation of killed or weakened microorganisms or viruses that is given to people or animals to generate a memory immune response.

ANTIGENIC DRIFT
Changes in viral antigens caused by genetic mutation during normal viral replication.

ANTIGENIC SHIFT
Changes in antigens that occur when viruses exchange genetic material with other strains.

Antigenic Drift and Shift Create New Influenza Strains

Mutation and gene exchange are two mechanisms by which influenza viruses can change. Mutations that accumulate gradually can cause variations in surface antigen. This process is called antigen drift. Different strains can also swap genes and cause surface antigens to change more dramatically. This is called antigenic shift. Drift is responsible for annual seasonal variation in influenza; shift is responsible for dramatic pandemics.

Antigenic Drift:
· Gradual change
· Caused by point mutations that occur when the virus replicates.

Antigenic Shift:
· Rapid change
· Caused by gene exchange between two different viruses that simultaneously infect the same cell.

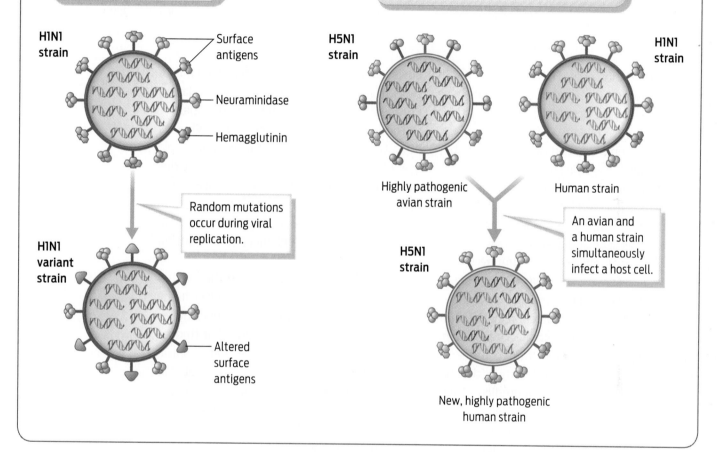

Severe acute respiratory syndrome (SARS), for example, is a respiratory illness caused by a virus. In 2003, more than 8,000 people worldwide became sick with SARS, according to the World Health Organization (WHO), and 774 died. Experts think the virus jumped from animals to people, although they have not yet determined the identity of the animal host.

They do, however, suspect that the source was a species of bird. Birds are a primary source of potentially pandemic viruses. Wild birds are routinely infected with viruses that live in their gastrointestinal tracts but cause little or no dis-

ease in the bird. Sometimes, however, bird flu viruses can jump from wild birds to poultry or other farm animals and become lethal. And since humans maintain such close contact with poultry, pigs, and other domestic animals, the risk of an animal flu infecting humans is high.

In the 1990s, a lethal strain of the avian flu virus (H5N1) crossed over into humans from domesticated birds, but it was not easily transferred from person to person. However, scientists continue to monitor this bird strain, not only because it can be deadly to the commercial poultry business, but also because the 1918 flu

TABLE 29.1

Known Flu Pandemics

NAME	DATE	VIRUS ANTIGEN SUBTYPE	NUMBER OF DEATHS
Asiatic (Russian) flu	1889–1890	H2N2	1 million
1918 (Spanish) flu	1918–1919	H1N1	20–100 million
Asian flu	1857–1958	H2N2	1–1.5 million
Hong Kong flu	1968–1969	H3N2	0.75–1 million
Swine flu	2009–present	H1N1	10,000 in 2009

carried a variation of the H5N1 bird flu genes. In fact, most human influenza pandemics have been caused by flu viruses that carry variations of bird flu genes (**Table 29.1**).

To help people ward off flu infection, public health officials offer flu shots every year. A flu shot is a vaccine: a weakened version or part of a flu virus is injected into the body in hope of generating a primary immune response and memory cells against that strain of virus. However, since influenza viruses mutate so frequently, a yearly flu shot may not protect us from getting the flu the following year. In fact, a flu shot may not confer protection for the duration of a season. Scientists create a new vaccine each year by tracking which strains of influenza are circulating worldwide and studying the antigens on their surfaces. But their predictions can be wrong. If public health officials decide to vaccinate against a strain of influenza with one variant of hemagglutinin antigen, but a strain with a different variant of the antigen is the one that strikes, even those who are vaccinated may become ill anyway. The specific memory cells they carry against flu won't protect them from a different strain of the virus.

This is one reason why public health officials are so concerned about the 2009 H1N1 influenza virus, or swine flu. H1N1 is a new virus strain that was first detected in people in the United States in April 2009 and became a worldwide pandemic in 2010. It was called swine flu because many of its genes were similar to those in influenza viruses that normally occur in North American pigs. Further study has shown, however, that this new virus is actually a mix of genes from flu viruses that normally circulate in European and Asian pigs, birds, and humans.

There are other reasons for concern. Whereas 90% of deaths from seasonal flu occur in people over 65 years old, H1N1 causes most severe disease in people under 25. This is the age group primarily affected by the 1918 virus, suggesting that swine flu, like Spanish flu, has the potential to become a more virulent infection that can kill within hours.

In 2009, Kawaoka reported in the journal *Nature* that the H1N1 virus could replicate in the lungs of mice, ferrets, and monkeys much better than a seasonal flu virus could. That finding "suggests the virus can cause serious respiratory illness in many people," Kawaoka stated in the paper. He also reported that some people born before 1918 have antibodies to the H1N1 virus, suggesting it shares some antigens in common with the 1918 virus.

Today, scientists have better surveillance, better communication, and can make more informed decisions about whether and how to quarantine groups to prevent a dangerous virus from spreading—even when extensive global travel might enable a new pandemic to spread rapidly.

In early 2010, the Bill and Melinda Gates Foundation awarded a $9.5 million 5-year grant to University of Wisconsin–Madison research scientists, who will be led by Kawaoka, to study viral mutations that could be early warning signs of pandemic flu viruses.

The 1918 flu pandemic taught scientists a lot about how viruses can become lethal in a matter of days, and demonstrated that developing vaccines can be like taking aim at a moving target. But by finding influenza genes common to strains that mutate less frequently than others, researchers may be able to develop a universal flu vaccine that would be effective over a number of years.

There are now at least two universal flu vaccines in clinical trials. Some companies are scaling up their vaccine manufacturing efforts, while others are boosting research and development of new universal vaccines. Some companies are also working on different antiviral drugs to help reduce the impact and spread of the influenza virus. The hope is that such measures will suppress the next big pandemic, whenever it happens, before it begins to rage. ■

How Do Other Organisms Defend Themselves?

Although not all organisms have as highly developed an immune system as do humans and other mammals, virtually all organisms have evolved ways to defend themselves from threats large and small.

Some organisms are able to repel invaders and predators by physical and chemical means. Sea sponges, for example, which represent early branches at the base of the evolutionary tree, are animals with different cell types but no distinct and differentiated tissues. They have no immune system, but they combine physical defenses with poisonous secretions to ward off danger. An internal support system made of collagen fibers and stiffened with hard spikes called crystalline spicules acts as a physical defense against predators like fish who may want to take a bite. In order to deter such predators from getting close enough to nibble, sponges also secrete chemicals that are toxic to many potential predators.

Many bacteria are able to protect themselves from a class of viruses known as bacteriophage–"phage" for short. Phage infection causes a bacterial cell to burst, or lyse–a life-ending event for that bacterium. But bacteria have defenses. They have restriction enzymes that act like scissors, cutting up any DNA that a phage injects into the bacterial cell. By destroying the phage DNA, bacteria prevent phages from replicating. At the same time, bacteria's own DNA is protected from these powerful molecular scissors.

Sea Sponges Employ Physical and Chemical Defenses

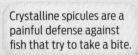

Crystalline spicules are a painful defense against fish that try to take a bite.

Toxins released by sponges keep predators from getting too close.

Bacterial cells that try to infect a sponge are lysed by molecules that punch holes in their cell walls.

They have a mechanism to chemically modify their DNA, which ensures their DNA won't be cut up by their own restriction enzymes.

Not all organisms that are evolutionarily ancient lack an immune system. Sea stars, for example, have a primitive immune system that consists of specialized cells that can attack or neutralize invaders. It has long been known that sea stars very rarely develop bacterial infection. It turns out that their ability to resist bacterial pathogens comes from a class of phagocytic cells called amoebocytes, which act much like the macrophages of the mammalian innate immune system. Amoebocytes circulate in the body cavity fluid of sea stars. When bacteria are injected into the body cavity of sea stars, the vast majority of amoebocytes actively

Bacteria Use Enzymes to Defend against Infection

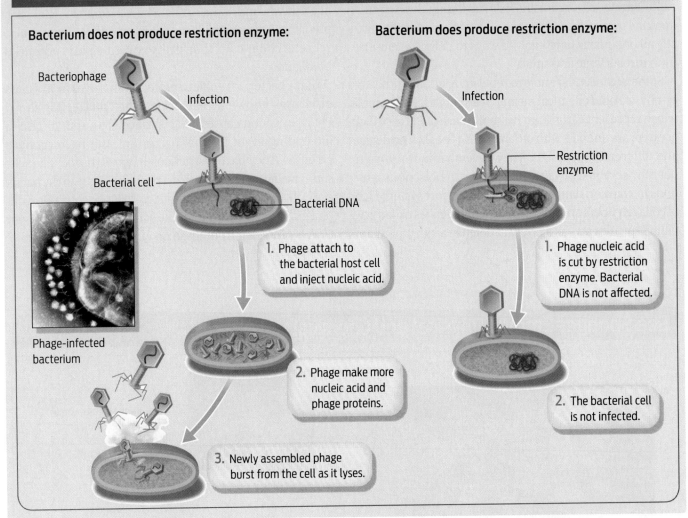

Bacterium does not produce restriction enzyme:

Bacteriophage

Infection

Bacterial cell

Bacterial DNA

Phage-infected bacterium

1. Phage attach to the bacterial host cell and inject nucleic acid.

2. Phage make more nucleic acid and phage proteins.

3. Newly assembled phage burst from the cell as it lyses.

Bacterium does produce restriction enzyme:

Infection

Restriction enzyme

1. Phage nucleic acid is cut by restriction enzyme. Bacterial DNA is not affected.

2. The bacterial cell is not infected.

engulfs and destroys the bacteria within 10 minutes. This rapid and efficient phagocytic response likely helps sea stars remain infection free.

Fruit flies also have several defense mechanisms that resemble the mammalian innate immune system. Phagocytic cells known as plasmatocytes act like human macrophages, ingesting and destroying foreign cells. Fruit flies also have cells called lamellocytes, which can coat and essentially wall off foreign objects, such as the eggs of parasitic wasps, that are too big to be phagocytosed. This walling off, or encapsulation, helps destroy the foreign object. Humans carry out a similar process in response to certain kinds of lung infections (including tuberculosis). In humans, immune cells surround and wall off the pathogen in a structure known as a granuloma. Thus, even some organisms that branched early on the evolutionary tree have defense mechanisms that are very much like ours.

Sea Stars Have Innate Cellular Defenses

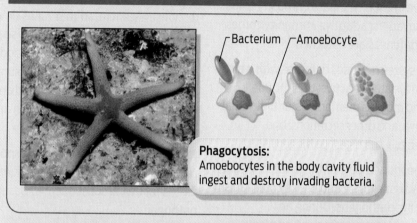

Bacterium — Amoebocyte

Phagocytosis:
Amoebocytes in the body cavity fluid ingest and destroy invading bacteria.

Fruit Flies Have Multiple Physical, Chemical, and Cellular Defenses

External Barriers:
Fruit flies have a hard exoskeleton and specialized cells that line the airways and digestive system.

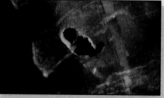

Encapsulation:
Lamellocytes (green) surround and wall off infecting parasites (black) inside living *Drosophila* larvae.

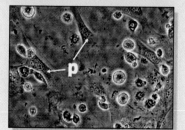

Phagocytosis:
Phagocytic cells known as plasmatocytes (labeled **p**) ingest and destroy invaders.

▶ Summary

■ The immune system defends the body against infection by pathogens.

■ Pathogens are particles that cause an immune response; they include certain bacteria, viruses, fungi, and parasites.

■ Viruses are noncellular; they consist of a genome of nucleic acid (DNA or RNA) contained within a protein shell.

■ The immune system has two components: innate immunity and adaptive immunity. Innate immunity is the first defense against invaders; adaptive immunity comes into play once innate defenses are breached.

■ The innate immune system includes defenses with which we are born, and which are always active: they include barriers such as skin and mucous membranes; antimicrobial chemicals in tears, saliva, and other secretions; and phagocytic cells that engulf and digest pathogens.

■ The inflammatory response, part of the innate immune system, is triggered by tissue injury and is a key mechanism for dealing with invading pathogens. During an inflammatory response, blood vessels swell and leak, marshaling phagocytic cells and protective molecules to the area to contain the infection.

■ Adaptive immunity is conferred by specialized lymphocytes called B and T cells.

■ B cells produce antibodies that specifically recognize antigens on pathogens and mark the pathogens for destruction.

■ T cells destroy infected cells and also stimulate B cells to produce antibodies.

■ The immune response can go awry, causing allergies, which are responses to intrinsically harmless substances (for example, pollen), and autoimmune conditions that result when the immune system mounts a response against the body's own cells and tissues.

■ At the first exposure to a particular pathogen, a primary response is generated that takes time to become fully effective; the primary response also produces memory cells. Memory cells remain in the body and become active at the time of a subsequent exposure to the same pathogen, producing a more rapid and vigorous secondary response that fights the pathogen and usually prevents the associated illness.

■ Vaccination elicits a primary response. Memory cells produced during the primary response protect against illness following subsequent exposure to the actual pathogen.

■ The adaptive response is highly specific for particular pathogens. If the pathogen changes, by antigenic shift or drift, the body will need to mount a separate response for each strain of the pathogen.

■ All organisms have ways of defending themselves against infection. Many organisms, including evolutionarily ancient ones, protect themselves by mechanisms similar to those of the human innate immune system.

VIRUSES AND OTHER PATHOGENS

Many viruses and bacteria can cause disease and are therefore pathogens.

HINT **See Infographics 29.1, 29.2, 29.10, and 29.11.**

➲ KNOW IT

1. Which of the following is found in all viruses?
   a. DNA
   b. RNA
   c. a membranous envelope
   d. a protein shell
   e. a cell membrane

2. Explain how viruses replicate within humans.

3. Why does poliovirus cause long-lasting damage, whereas those infected by influenza virus typically make a full recovery?

4. What is the difference between antigenic shift and antigenic drift?

➲ USE IT

5. Both viruses and bacteria can be human pathogens. Describe some key differences between them.

6. Why do poliovirus, influenza virus, and HIV infections cause different symptoms?

INNATE IMMUNITY

Innate immunity includes physical and chemical barriers and the inflammatory response, the first lines of defense of the body against pathogens.

HINT **See Infographics 29.3 and 29.4.**

➲ KNOW IT

7. Name three components of the innate immune system. For each, provide a brief description of how it offers protection.

8. What do macrophages and neutrophils have in common?

➲ USE IT

9. From what you know about innate immunity, would you predict different or identical innate responses to infections from *E. coli* (a bacterium) and *S. aureus* (another bacterium)? Explain your answer.

10. Neutropenia is a deficiency of neutrophils. Would you expect someone with neutropenia to be able to mount an effective inflammatory response? Explain your answer.

11. Why are those with influenza infections susceptible to bacterial pneumonia?

12. Why might those taking anti-inflammatory drugs be more susceptible than others to bacterial infections?

ADAPTIVE IMMUNITY

Adaptive immunity is a specific response to pathogens and other foreign antigens.

HINT **See Infographics 29.3 and 29.6–29.9.**

➲ KNOW IT

13. Compare and contrast the features of innate and adaptive immunity.

14. B cells, plasma cells, and antibodies are all related. Describe this relationship, using words, a diagram, or both.

➲ USE IT

15. Anti–hepatitis C antibodies in a patient's circulation indicate
   a. that the patient is mounting an innate response.
   b. that the patient has been exposed to HIV.
   c. that the patient has been exposed to hepatitis C within the last 24 hours.
   d. that the patient has been exposed to hepatitis C at least 2 weeks ago.
   e. that the patient has hepatitis.

16. Vaccination against a particular pathogen stimulates what type of response?

    a. innate

    b. primary

    c. secondary

    d. autoimmune

    e. b and c

17. Will someone who has been exposed to seasonal influenza in the past

    a. have memory B cells?

    b. still be at risk for seasonal influenza next year? Why?

    c. still be at risk for H1N1 (swine flu)? Why?

18. What processes are responsible for the emergence of pandemic influenza strains, such as H1N1 swine flu? Explain how these strains can spread so successfully through the human population.

19. *Staphylococcus aureus* causes a bacterial skin infection that can become very serious.

    a. Why does the body exhibit innate and adaptive responses to *Staphylococcus aureus* but not to its own skin cells?

    b. Will the innate response to *Staphylococcus aureus* be equally effective against *Streptococcus pyogenes*, another bacterium that can cause skin infections? Explain your answer.

    c. Will the adaptive response to *Staphylococcus aureus* be equally effective against *Streptococcus pyogenes*? Explain your answer.

20. HIV is a virus that infects and eventually destroys helper T cells. Why do people with AIDS (advanced HIV infections) often die from infections by other pathogens?

SCIENCE AND SOCIETY

21. Vaccination against a potentially pandemic disease, such as measles or influenza, lessens the probability of developing disease. Would it be a good idea for an agency such as a local school board to require that everyone in its jurisdiction be vaccinated?

22. What are the relative merits of investing in disease prevention (for example, by vaccines) or in disease treatment (for example, by antibiotics and antivirals)?

Q & A: Plants

Bugs are drawn to what look like dewdrops on this carnivorous plant, the Australian sundew. The "dewdrops" turn out to be sticky mucus bubbles that trap the insects so the plant can digest them.

Q & A: Plants

Plants have evolved a unique set of solutions to nature's challenges

Like other organisms, plants are products of evolution. They are equipped with adaptations that help them survive and flourish in their particular environments. Plants face many of the same life challenges as do other organisms—obtaining nutrients, reproducing, protecting themselves—but their solutions to these challenges are often unique and surprising. We encountered the evolution of plant diversity in Chapter 19; here we focus on plants that have evolved to succeed on dry land.

ROOT SYSTEM
The belowground parts of a plant, which anchor it and absorb water and nutrients.

SHOOT SYSTEM
The aboveground parts of a plant, including the stem and photosynthetic leaves.

PLANT STRUCTURE

Q Plants don't have bones—so how do they stand up?

A From tiny 3-inch-tall crocus flowers to massive redwood trees standing 300 feet, nearly all land plants share the same basic design: a belowground **root system** for absorbing water and nutrients and an aboveground **shoot system** made up of stems and leaves. Bones not included. In the absence of a bony skeleton, what keeps a plant from flopping over?

Like other eukaryotes, plants are made of cells packed with organelles, including a nucleus, endoplasmic reticulum, and mitochondria. Plant cells contain a few plant-specific parts as well, including a supportive **cell wall; chloroplasts,** the sites of photosynthesis; and a **central vacuole,** essentially a large water balloon occupying the center of the cell. When filled with water, vacuoles create **turgor pressure** against the cell wall, keeping a plant body rigid and upright. When the vacuoles are less than full, turgor pressure is reduced, and the plant wilts.

Cell walls contribute to a plant's stiffness in another way. Plant cells are packed tightly together, much like bricks in a wall; carbohydrates that make up the cell walls act like glue, helping adjacent cells stick together. With many cells held together in this way, plant tissues are exceptionally strong, which is why they make such durable ropes and fabrics.

Some plant tissues, such as those that make up the stem, are made of cells with two cell walls: an outer cell wall containing cellulose, and a second, inner cell wall containing cellulose and **lignin.** Lignin is a hard, durable material, which lends added support and strength to plant tissues; it is what makes them "woody." A plant with a thick, woody stem (or trunk, in the case of a tree) can grow tall and still not topple over. The tallest of all plants is the

> **The tallest of all plants is the California redwood (*Sequoia sempervirens*), which can reach heights of more than 350 feet.**

California redwood (*Sequoia sempervirens*), which can reach heights of more than 350 feet.

Plants would not be able to grow so tall were it not for the extensive system of roots that anchors them in the ground. Some plants have root systems that are much deeper than the plants are tall, and some plants' roots extend out horizontally to about three times the branch spread. A vast number of tiny **root hairs** increase the surface area of the root, enhancing its ability to absorb water and nutrients. When root hairs are included, a plant's total root length can be hundreds of miles long. In addition, some plants produce very long **taproots**—deep-reaching roots that help the plant reach the underground water table. Taproots also store water and carbohydrates for the plant, making them appealing and nutritious to animals. Carrots and turnips are actually taproots **(Infographic 30.1).**

Q Why don't plants bleed?

A Like animals, plants have a **vascular system** for transporting valuable fluids. Instead of blood, however, a plant's vascular system transports nutrients like water and sugar throughout the plant body. Plants need water to live and grow (you may have learned this the hard way if you've ever killed a plant with neglect). Among other things, water is crucial for photosynthesis in leaves. But you don't water a plant's leaves; you water the roots by pouring water in the soil. How does water get from the roots to the rest of the plant?

A plant's water-carrying tissue is known as **xylem** (pronounced ZYE-lum). Xylem tissue is made of cells arranged into long, stiff tubes; the cells have holes in each end and are stacked one on top of the next. Water moves up from the roots through these tubes to the aboveground stems, and eventually into the leaves, where it is used during photosynthesis to make sugar (for more on photosynthesis, see Chapter 5). Plant

All Plants Share the Same Basic Design

Trees and other plants are anchored in the soil by a root system and extend their shoot system, including stems and leaves, skyward. A variety of features, including rigid cell walls and a central vacuole, allow the stems to remain upright.

Leaves
The site of photosynthesis

Stem
Rigid, upright support for the plant

Roots
Anchor the plant and absorb water and minerals from the soil

Taproot
Extends deep into the ground to anchor plants and reach underground sources of water

Plant Cell

Chloroplast
Photosynthetic organelle in leaf cells

Nucleus
Compartment for DNA

Cell Wall
Rigid cellulose-based structure surrounding the cell membrane

Central Vacuole
Fills with water, creating turgor pressure against the rigid cell wall and thus keeps a plant from wilting

Root Hairs
Extensions on root cells that increase the surface area for enhanced absorption of water and minerals from the soil

cells use the newly synthesized sugar as food, so the sugar must be transported out of the leaves and back down through the plant. Another series of tubes, known as **phloem** (pronounced FLO-um; think "f" for "food"), transports sugar through the plant. Phloem supports two-way transport: sugar moves down to the roots, where some of it is stored; later, sugar moves up to the shoot system, where it provides nutrition and energy for growing fruit, buds, and leaves.

> **Xylem and phloem, like the vessels of the human circulatory system, transport all the essential nutrients the plant body needs to live and grow.**

Xylem and phloem, like the vessels of the human circulatory system, transport all the essential nutrients the plant body needs to live and grow.

Moving water and other nutrients up to the top of a 300-foot tree against the force of gravity is no easy task–hundreds of pounds of water must be lifted up through what is essentially a long water pipe. And unlike animals, which have a heart to pump blood along, plants have no such mechanical help. Instead, plants rely on evaporation of water from the

PHLOEM
Plant vascular tissue that transports sugars throughout the plant.

TRANSPIRATION
The loss of water from plants by evaporation, which powers the transport of water and nutrients through a plant's vascular system.

leaves to siphon up water in a process called **transpiration.** Because water is a polar molecule that can form hydrogen bonds with other water molecules, it has great cohesive strength (see Chapter 2). As water evaporates from leaves into the air, water in the xylem is pulled up to replace it. The cohesive strength of water is enough to counteract the force of gravity and pull water up to astonishing heights.

For a plant, transpiration is life sustaining: it is the force that carries water through the plant. But losing too much water can be dangerous. On a hot, dry, or windy day, a large tree can lose hundreds of gallons of water vapor from its leaves. To control the amount of water lost by transpiration, a plant's leaves are coated with a waxy **cuticle** that functions somewhat like a rubber suit, sealing in moisture. At regular intervals, the cuticle is punctuated by pores, called **stomata,** which open and close. When stomata are open, water vapor leaves freely and other gases enter and exit–specifically, carbon dioxide and oxygen. When stomata are closed, water and gases are sealed in. Many plants keep their stomata open during the day to let in carbon dioxide for photosynthesis. At night, they close the stomata to conserve water **(Infographic 30.2).**

INFOGRAPHIC 30.2

Plants Transport Water and Sugar through Their Vascular System

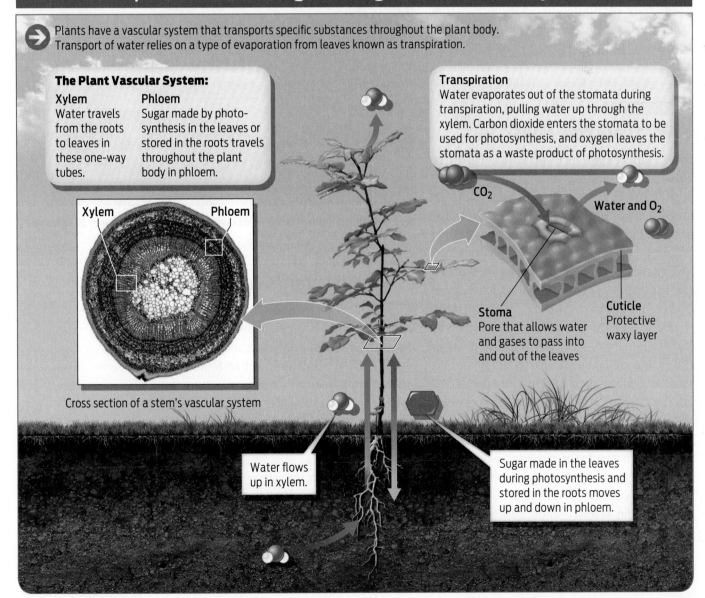

Plants have a vascular system that transports specific substances throughout the plant body. Transport of water relies on a type of evaporation from leaves known as transpiration.

The Plant Vascular System:

Xylem
Water travels from the roots to leaves in these one-way tubes.

Phloem
Sugar made by photosynthesis in the leaves or stored in the roots travels throughout the plant body in phloem.

Xylem   Phloem

Cross section of a stem's vascular system

Transpiration
Water evaporates out of the stomata during transpiration, pulling water up through the xylem. Carbon dioxide enters the stomata to be used for photosynthesis, and oxygen leaves the stomata as a waste product of photosynthesis.

CO_2

Water and O_2

Stoma
Pore that allows water and gases to pass into and out of the leaves

Cuticle
Protective waxy layer

Water flows up in xylem.

Sugar made in the leaves during photosynthesis and stored in the roots moves up and down in phloem.

Eventually, if a plant loses too much water, it will die. During an extreme drought, if you put a stethoscope to a tree and listen closely, you may hear clicks. This is the sound of the rigid column of water in the xylem breaking because there is no water in the soil to replace the water being lost by transpiration.

The presence of specialized tissues for transporting water is what distinguishes plants from their water-dwelling ancestors, the algae. The evolution of this vascular tissue is what allowed plants to colonize nearly every part of the land, from valley to mountaintop. A few primitive nonvascular land plants, such as mosses and liverworts, also lack true roots and shoots containing vascular tissue. Without these specialized tubes for transporting water, they are limited to environments that are saturated with water, where they grow close to the ground in squat, spongy mats.

Each dark ring represents the end of 1 year of growth. This tree was 37 years old.

PLANT GROWTH

Q What are tree rings?

A Plants experience two kinds of growth: primary and secondary. Primary growth is growth in length–a tree getting taller. Trees and other plants also grow wider as well–this is secondary growth. Tree trunks increase in diameter because they add a new layer of xylem tissue–otherwise known as **wood**– each year. In temperate regions, such as most of the United States, xylem growth is dormant in the winter. In the spring, xylem growth starts again. The first xylem cells to grow in the spring are usually larger in diameter and thinner-walled than xylem cells produced later in the summer. The boundary between the smaller xylem cells from one summer and the larger xylem cells from the next spring appears as a ring in the cross section of a tree trunk. You can determine the age of a tree by counting the number of rings. Tree rings vary in width from year to year because differences in temperature and rainfall affect the amount of xylem tissue that is produced in any one season.

PLANT NUTRITION

Q What do plants eat?

A This is a a trick question, because plants don't "eat" in the same sense as animals. Plants are autotrophs, meaning they make their own food from inorganic materials found in the environment. Give a plant some sunshine, carbon dioxide, and water–plus a few soil nutrients–and you will have a happy plant, capable of feeding and nourishing itself. From the air, plant leaves absorb carbon dioxide, which they convert into sugar through photosynthesis. Sugar is a source of energy and building blocks for the plant. From the soil, plant roots absorb nutrients–primarily nitrogen– that are required to make plant proteins and other molecules.

WOOD
Secondary xylem tissue found in the stem of a plant.

Although nitrogen gas is plentiful in air spaces in the soil, it exists in a form that is essentially unusable by plants. Luckily, bacteria within the soil are able to convert nitrogen into a form that can be used by plants, a process known as **nitrogen fixation** (for more, see Chapter 23). Some of these bacteria, like members of the genus *Rhizobium,* can even live symbiotically in a plant's roots, forming lumpy structures known as root nodules on the roots of legumes. By making plant growth possible, nitrogen-fixing bacteria play a critical role in supporting nearly all life on the planet.

Fertile soil naturally contains nitrogen-fixing bacteria and therefore provides adequate supplies of usable nitrogen for plants to absorb and use to grow. And soil can be supplemented with fertilizer, giving plants a boost of artificial nitrogen and other nutrients. But in certain natural environments, such as bogs or rock outcroppings, it's hard for plants to obtain the nitrogen they need. The acidity of a bog, for example, prevents organic matter from breaking down, so nutrients are recycled more slowly. In these environments, plants have evolved novel ways to obtain scarce nitrogen–some of which would put animal carnivores to shame.

Trumpet pitchers (*Sarracenia*), for example, lure insects with brightly colored flowers and nectar "bribes." But the rim of the plant's trumpet-shaped flower is slippery. Unsuspecting trespassers climb onto the rim, lose their grip, and tumble into a deep cavity filled with digestive juices. Prevented from escape by downward-pointing spikes, the tiny prisoners drown and are slowly dissolved. The resulting insect soup–a rich source of nitrogen–is absorbed by the plant.

The Venus flytrap (*Dionaea muscipula*) takes an even more dramatic approach. The plant's "flower" is actually a spring-loaded trap that snaps shut around unsuspecting prey. Tiny hairs inside the flower act as sensors; when the sensors are tripped by a moving insect, the trap slams shut and the feasting begins **(Infographic 30.3)**.

Even when plants obtain nitrogen in this "carnivorous" way, they must still perform photosynthesis to make sugar. The plant body is composed of complex carbohydrates, such as cellulose, which the plant makes by stringing sugar molecules together. And the starting material for photosynthesis is carbon dioxide gas. Thus the air, rather than the soil, is where a plant obtains the material to put on weight.

Ⓠ Can plants photosynthesize at night?

Ⓐ By definition, there is a crucial part of photosynthesis that can occur only during daylight hours–the light-absorbing "photo" part. So technically, the answer is, no–plants can't photosynthesize at night. However, some plants can collect carbon dioxide in the dark, a useful capability in hot, dry climates.

Because daylight hours tend to be the time of day when it's hottest and driest, plants lose a lot of water during the day. To conserve water, plants can close their stomata. But closing stomata also prevents carbon dioxide–another crucial ingredient for photosynthesis–from entering the leaf. In many plants, wheat and rice, for example, the result of this trade-off is reduced output of plant food: sugar.

Some plants have adapted to sun-scorched surroundings. Corn and sugar cane, for example, are able to thrive in hot, sunny climates by keeping their stomata closed as much as possible. They use a molecule known as PEP carboxylase to capture CO_2 even when it is present only at low concentrations in air pockets in the leaf when the stomata are closed. Once the CO_2 has been captured, it can be "fed" into the synthesis part of photosynthesis to generate the sugar that the plants rely on.

Still other plants have adapted to hot, dry conditions by splitting up two parts of photosynthesis that usually occur simultaneously: taking in CO_2 and making sugar. Pineapples as well as cacti and other succulent plants–juicy aloe and jade plants, for instance–conserve water by keeping their stomata closed during the day when it's hot. But at night, when it's cooler, they open their stomata to allow carbon dioxide in. The CO_2 that is captured at night isn't

NITROGEN FIXATION
The process of converting atmospheric nitrogen into a form that plants can use to grow.

Plants Obtain Nutrients in a Variety of Ways

 Plants obtain carbon from atmospheric CO_2 during photosynthesis. They obtain minerals and nutrients such as nitrogen from the soil, often with a little "supplementation."

Plants are autotrophs: they make their own food through photosynthesis, with atmospheric CO_2 providing the carbon to build molecules such as sugars.

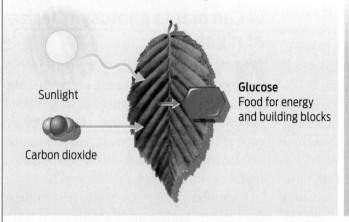

Sunlight

Carbon dioxide

Glucose
Food for energy
and building blocks

Plants get other nutrients necessary for plant growth from the soil.

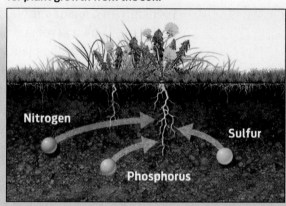

Nitrogen

Sulfur

Phosphorus

Carnivorous plants in nutrient-poor soil obtain nutrients such as nitrogen by digesting insects.

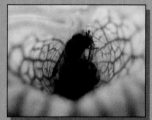

Trumpet pitcher plants attract insects with bright colors and sweet nectar.

Fly trapped inside a trumpet pitcher plant

Some plants get their nitrogen by "partnering" with nitrogen-fixing bacteria that live in association with their roots.

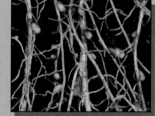

Rhizobium
A nitrogen-fixing bacterium

Clover plant roots with nodules containing nitrogen-fixing bacteria

used right away but instead is stored. During the day, these plants can leave their stomata closed and use the stored CO_2 to complete photosynthesis and make sugar. By segregating in time the steps of photosynthesis, these well-adapted plants have found a way to thrive in conditions that would wither their less physiologically adapted cousins **(Infographic 30.4)**.

Like other eukaryotic organisms, plants use sugar to perform aerobic respiration to make ATP (Chapter 6). All plants respire both during the day and at night. Because the total amount of carbon dioxide given off by plants during cellular respiration is less than the total amount taken in for photosynthesis, plants are carbon dioxide sinks that ultimately lower the amount of carbon dioxide in the atmosphere.

Q Why do leaves change color in the fall?

A For most of the year, leaves are photosynthesis factories, using sunlight, water, and carbon dioxide to make sugar. A key component of the photosynthetic machinery is a pigment called **chlorophyll,** which absorbs red and

CHLOROPHYLL
The dominant pigment in photosynthesis, which makes plants appear green.

INFOGRAPHIC 30.4

Beating the Heat: Some Plants Conserve Water

→ All plants must perform the "photo" reactions of photosynthesis during the day. But many plants experience reduced levels of photosynthesis in hot and dry conditions. Other plants are adapted to live in hot and dry climates, and have different strategies to take in CO_2 while minimizing water loss.

Stomata open all day: too much water loss

CO_2 in

H_2O out

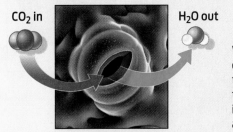

When stomata are open, carbon dioxide can enter the leaves for photosynthesis. However, this increases transpiration and therefore water loss.

Stomata closed all day: too little sugar produced

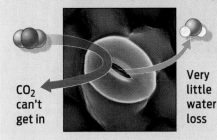

CO_2 can't get in

Very little water loss

In dry climates, some plants keep stomata closed during the hot hours of the day. This conserves water, but inhibits the uptake of carbon dioxide for photosynthesis. These plants make less sugar food.

Aloe ferox

To combat this problem, some plants in dry climates:

1. Conserve water by having thick, waxy succulent leaves.
2. Open stomata at night to capture CO_2, while reducing water loss by transpiration.
3. Extract CO_2 from airspaces within the plant even when stomata are mostly closed to conserve H_2O.

blue wavelengths of light and reflects green wavelengths. Chlorophyll is the reason plants appear green. To keep photosynthesis running, plants make abundant chlorophyll in spring and summer. But they also produce a smaller amount of other pigments—yellow-reflecting xanthophyll and orange-reflecting carotene. In the leaves, these pigments capture additional wavelengths of light and therefore expand the range of light that is effective for photosynthesis. You can't see these other pigments in leaves during spring and summer because leaves are chock full of green chlorophyll, camouflaging the other hues (although you can see them elsewhere in some plants: in the flesh of a pineapple, for example, or the root that is a carrot).

After a summer of intense sugar stockpiling, trees, bushes, and other deciduous plants that seasonally drop their leaves start to settle in for the winter and begin to shut down their photosynthesis machinery. During the winter months in temperate regions, there isn't enough sunlight or water to drive photosynthesis; days are shorter and the water in the ground is frozen and can't be absorbed. As temperatures cool and daylight wanes, plants turn off the production of their light-absorbing pigments. Of all the pigments, chlorophyll is the most chemically unstable and therefore the most short lived: levels fall quickly once production stops. By contrast, xanthophyll and carotene linger longer. As green fades, the other colors peak through—mostly yellow and orange, which have been there all along, but hidden.

The really intense colors—the fiery reds and deep purples—that some trees and bushes turn in autumn are the result of a fourth pigment, called anthocyanin. This distinctive chemical, which is

Plants Produce Multiple Light-Capturing Pigments

 Leaves contain chlorophyll and other pigments. These pigments help capture a wide range of wavelengths of light to maximize the efficiency of photosynthesis.

More Chlorophyll:

Green leaves produce an abundance of the green pigment chlorophyll during the warm, sunny months. They also produce smaller amounts of red and yellow pigments.

Less Chlorophyll:

When the hours of daylight are shorter and the temperature is cooler, the trees perform less photosynthesis and do not produce chlorophyll. The red, purple, and yellow pigments produced by the leaves are revealed.

the same one that gives apples and acai berries their color, is produced in leaves in response to high sugar concentration. Why does sugar concentration rise in the fall? As trees prepare to lose their leaves in winter, a corklike membrane develops between the leaf stem and the branch, sealing off the leaf and preventing phloem from transporting sugar out of the leaf. Unable to move out of the leaf, sugar begins to collect. The more sunny days there are in fall, the more photosynthesis occurs, the more sugar is created and trapped in the leaves. More sugar means more anthocyanin is produced, yielding the bright colors we associate with fall.

Some trees naturally produce lots of anthocyanin—maple and oak, for example—which explains why these trees produce such brilliant colors. Others, such as aspen and poplar, produce less anthocyanin and turn yellow or orange in fall **(Infographic 30.5)**.

Eventually the corky membrane that seals off the leaf from the tree causes the leaf to dry out. With a slight gust of wind, the leaf flutters to the ground. By shedding its leaves, a tree protects itself from water loss by transpiration in cold, dry air. The tree is now prepared for winter.

PLANT REPRODUCTION

Q Do plants have sex?

A They may not go on blind dates or place personal ads, but plants are among the most "flirtatious" and sexually active organisms on earth. The many bright colors, provocative shapes, and alluring fragrances of flowers are the plant equivalents of lipstick, muscle shirts, and perfume—evolutionary novelties designed to attract pollinating suitors.

Like other sexually reproducing creatures, plants have distinct male and female reproductive

STAMEN
The male sexual organ of a flower.

POLLEN
Small, thick-walled structures that contain cells that will develop into sperm.

PISTIL
The female reproductive organ of a flower.

POLLINATION
The transfer of pollen from a male stamen to a female pistil.

organs that produce haploid gametes, sperm and egg. In flowering plants, male and female sexual organs are housed within a plant's flowers. The male sexual organ, called a **stamen,** consists of a stalklike filament that supports a structure, the anther, that contains **pollen.** Pollen grains contain cells that will develop into sperm. The female structure, called a **pistil,** consists of a tubelike style topped with a sticky stigma, a "landing pad" for pollen. At the base of the pistil is a plump ovary stuffed with egg-containing ovules. Transfer of pollen from a male stamen to a female pistil results in **pollination,** which may then lead to fertilization. To fertilize eggs, sperm from the pollen must travel down the pistil to the ovule, where the eggs are located. When a haploid sperm fuses with a haploid egg, the result is a diploid plant embryo.

Plants are among the most "flirtatious" and sexually active organisms on earth.

In some flowering plant species, male and female organs are found on separate male and female plants. Ginkgo trees and holly bushes are examples of such unisexual plants. More commonly, however, flowering plants–from dandelions to lilies–are hermaphrodites: their flowers contain both male and female reproductive parts. Fittingly, a few such hermaphroditic plants can self-pollinate–transferring pollen from stamen to pistil within the same flower or plant (Mendel's pea plants could do this). But most hermaphroditic plants have evolved ways to prevent self-pollination and to encourage outcrossing instead. They may have pistils and stamens of different heights, for example, or their pollen may be chemically rejected by eggs of the same plant.

Plants Reproduce Sexually

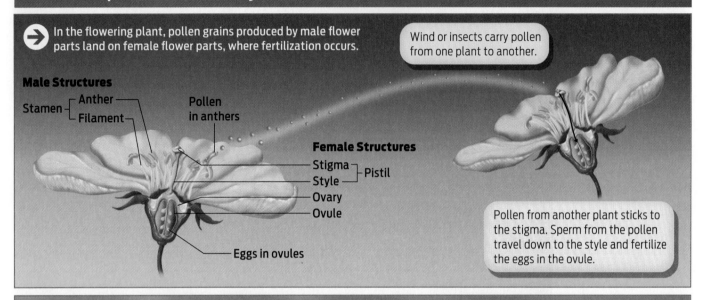

→ In the flowering plant, pollen grains produced by male flower parts land on female flower parts, where fertilization occurs.

Wind or insects carry pollen from one plant to another.

Male Structures

Stamen — Anther
         — Filament

Pollen in anthers

Female Structures

Stigma ⎤
Style ⎦ Pistil
Ovary
Ovule

Eggs in ovules

Pollen from another plant sticks to the stigma. Sperm from the pollen travel down to the style and fertilize the eggs in the ovule.

Birds and bats are also essential pollinators for many flowering plants.

Plants have evolved many elaborate ways to spread their pollen from plant to plant. For many flowering plants, pollinators such as bees and hummingbirds transfer pollen grains from the anther of one flower to the stigma of another. Other plants rely on wind to transfer pollen (Infographic 30.6).

Not all plants are so sexually adventurous. Some are asexual, or have asexual phases, and can create new individuals through underground root runners or bulbs that develop into whole new plants. Asexual reproduction, in the form of cuttings and grafting, is especially important in agriculture. A cutting taken from one plant can be planted in soil to form a new plant; sugar cane and pineapples are often reproduced this way. A cutting taken from a plant can also be grafted to the root system of a different plant—a procedure commonly used to perpetuate vineyard grapes.

Q What spins like a helicopter, shoots like a rocket, and contains its own parachute?

A The answer to this biological riddle: a seed. Unlike animals, land plants cannot move to seek out more hospitable living conditions for their offspring when resources are scarce or the neighborhood gets too crowded. Forest or desert, valley or mountaintop, plants are stuck where nature put them. Their solution to this enforced sedentary existence is to disperse their offspring far and wide through seeds. A **seed** is a small embryonic plant contained within a sac of stored nutrients–essentially, a plant starter kit. It develops from a fertilized egg and is a perfect package for delivering an immature plant to its new home. It's also a tremendously successful evolutionary adaptation, which is why two types of seed-bearing plants dominate the plant world: **gymnosperms,** which include seed-cone-producing conifers, and **angiosperms,** the flowering plants. "Gymnos" is Greek for "naked," so the name literally means "naked seeds"; in a pinecone, for example, seeds sit nestled under the cone's scales. In angiosperms, seeds are located inside a fruit or nut: "angio" is derived from the Greek for "vessel" or "container." Angiosperms make up nearly 90% of all plants–including most of our agricultural crops (see Chapter 22).

There are many shapes and sizes of seeds, and they are dispersed in myriad ways. Some are small and hitch a ride on fur or clothing by means of tiny hooks or burrs. Some, like coconuts, are large and buoyant and can float across oceans to reach distant beaches. The delicate parachutes of dandelions, the spinning helicopters of maples and pine trees, and other lightweight seeds sail on the wind. Cottonseeds are little more than hairy specks of dirt, but on a steady breeze they can windsurf for miles. Other seeds are packaged inside fruit: tempted by its bright color and sugary content, animals eat the fruit and then deposit the seeds in feces some distance away from the original plant. Some seeds are dispersed through a ballistic mechanism. The seedpod of the squirting cucumber (*Ecballium elaterium*), for example, fills with slimy juice as it ripens. Eventually, the mounting pressure of

Land plants cannot move to seek out more hospitable living conditions for their offspring when resources are scarce.

the increased volume of juice causes the cucumber to shoot off the plant like a rocket, trailing a plume of seeds and slime in its wake. At the slightest touch, the bulging seedpods of the aptly named touch-me-not plant explode, spraying seeds like bullets. When a seed lands in favorable conditions, with enough water, it will germinate and grow into a young plant, or seedling **(Infographic 30.7)**.

PLANT HORMONES

Q Can plants see?

A Observe an old building and you'll likely see ivy scaling up its walls. Keep houseplants next to a window and you'll find them bending toward the sunlight. How does a plant know where it's going? While plants don't have eyes and therefore cannot see, they are quite adept at sensing and responding to their environment, which they do through various kinds of tropism (from the Greek "tropos," meaning "turn").

The growth of a plant shoot toward light is called **phototropism.** This is how leaves get the sunlight they need for photosynthesis, and it's why ivy climbs up walls and houseplants lean into the window. Shoots grow towards the light because of **auxin,** a plant hormone that promotes cell elongation as one of its effects. When light hits one side of a plant shoot, auxin moves to the shaded side, creating a gradient of the hormone in the stem. The side receiving the most direct sunlight contains the least auxin, while the shaded side contain the most. Auxin promotes elongation of cells on the shady side of the stem. This causes the shaded side to elongate faster than the sunny side, pushing the stem toward the sun. The whole stem doesn't sense light and produce auxin, though–just the tip of the stem. Cover the tip of a young plant and it won't turn toward the light, but will instead grow straight up because, in the absence of

SEED
An embryonic plant contained in a protective structure.

GYMNOSPERMS
Cone-bearing seed plants.

ANGIOSPERMS
Flowering plants.

PHOTOTROPISM
The growth of the stem of a plant towards light.

AUXIN
A plant hormone that causes elongation of cells as one of its effects.

Seeds Carry a Young Plant to a New Destination

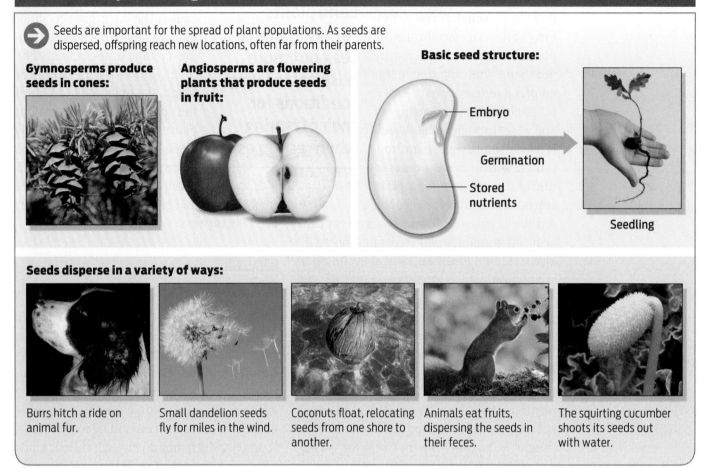

→ Seeds are important for the spread of plant populations. As seeds are dispersed, offspring reach new locations, often far from their parents.

Gymnosperms produce seeds in cones:

Angiosperms are flowering plants that produce seeds in fruit:

Basic seed structure:

Embryo

Germination

Stored nutrients

Seedling

Seeds disperse in a variety of ways:

Burrs hitch a ride on animal fur.

Small dandelion seeds fly for miles in the wind.

Coconuts float, relocating seeds from one shore to another.

Animals eat fruits, dispersing the seeds in their feces.

The squirting cucumber shoots its seeds out with water.

---

sunlight, auxin is not differentially localized to one side of the stem.

Other mechanisms help a plant sense where it is and orient itself in space. **Gravitropism** is the growth of plants in response to gravity–roots grow downward, with the force of gravity, and shoots grow upward, against it. Auxin is again the main player in this mechanism. When a plant is placed on its side, more auxin is sent to the down side of the stem, in the direction of gravity. This causes the cells on the down side to elongate more. Stems begin to curve away from gravity. Root cells, however, respond the opposite way to auxin: more auxin on the gravity side of roots inhibits root cell elongation on the down side, so roots bend toward gravity. Gravitropism allows a planted seed to send its shoots toward the light and its roots toward the soil. The same chemical signal–auxin–is able to produce these opposite effects because different tissues respond differently to the hormone.

A plant's sense of touch is called **thigmotropism;** it's how vines sense their way around poles or trellises and carnivorous

An example of gravitropism in the remains of a cellar in a Roman villa in the Archeologic Park in Baia, Italy. The leaf-bearing stems of this upside-down tree bend upward away from gravity.

GRAVITROPISM
The growth of plants in response to gravity. Roots grow downward, with gravity; shoots grow upward, against gravity.

THIGMOTROPISM
The response of plants to touch and wind.

ETHYLENE
A gaseous plant hormone that promotes fruit ripening as one of its effects.

plants sense their prey. Touch-sensitive growth is also controlled by auxin. Vine cells touching a pole, for example, get less auxin and consequently elongate less, while cells on the other side elongate more. The result is another kind of lopsided growth in the shoot, which eventually causes the vine to coil around whatever it's touching. A plant's sense of touch can be exquisitely sensitive–more sensitive than a human's. A human can detect the presence of a thread weighing 0.002 mg laid across the arm. By contrast, the feeding tentacle of the insectivorous sundew plant can sense a thread of less than half that weight. The legs of a single gnat are enough to trigger the tentacle into swift action **(Infographic 30.8)**.

Does one bad apple really spoil the bunch?

A This old adage is true, and its truth can be shown empirically: put a ripe apple in a bowl of unripe ones, and the unripe neighbors will quickly ripen. That's because ripe fruit–bananas and apples, especially–produce **ethylene,** a gaseous plant hormone, one effect of which is to promote ripening. In a confined space, the ethylene gas collects and causes nearby fruit to ripen through the loss of chlorophyll and the breakdown of cell walls. The result is the conversion of a hard, green fruit to a soft, ripe one. Commercial fruit growers take advantage of the action of ethylene when they ship fruit

INFOGRAPHIC 30.8

Plants Sense and Respond to Their Environment

→ Plants can respond to variety of stimuli, including light, touch, and gravity. Many of these responses are mediated by a plant hormone called auxin. Unequal distribution of auxin leads to a corresponding unequal pattern of cell elongation in stems and roots, and thus growth in a particular direction.

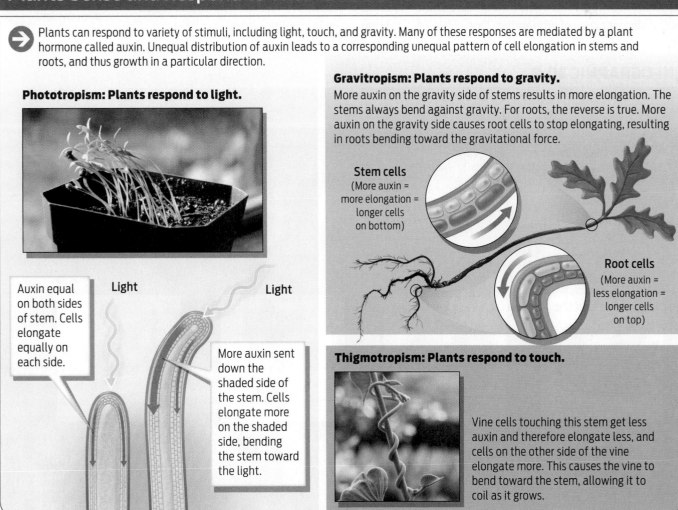

Phototropism: Plants respond to light.

Auxin equal on both sides of stem. Cells elongate equally on each side.

Light

Light

More auxin sent down the shaded side of the stem. Cells elongate more on the shaded side, bending the stem toward the light.

Gravitropism: Plants respond to gravity.

More auxin on the gravity side of stems results in more elongation. The stems always bend against gravity. For roots, the reverse is true. More auxin on the gravity side causes root cells to stop elongating, resulting in roots bending toward the gravitational force.

Stem cells
(More auxin = more elongation = longer cells on bottom)

Root cells
(More auxin = less elongation = longer cells on top)

Thigmotropism: Plants respond to touch.

Vine cells touching this stem get less auxin and therefore elongate less, and cells on the other side of the vine elongate more. This causes the vine to bend toward the stem, allowing it to coil as it grows.

to distributors. Often, fruit is picked while still green and then exposed to natural or synthetic ethylene just before arrival in grocery stores to hasten ripening. This method works especially well with tomatoes, avocados, bananas, and cantaloupe. In some cases, fruit growers remove ethylene from storage containers in order to prevent ripening so that fruits can be stored for a long time—apples that are picked in fall and sold in summer, for example (Infographic 30.9).

As a plant hormone, ethylene does more than ripen fruit. It is responsible for a number of different aging effects in plants—including leaf dropping in autumn.

Q Can plants take hormones to improve their performance?

A In a manner of speaking, yes. **Gibberellins** are plant hormones that promote growth. Scientists have identified more than 100 different types of gibberellin. One effect of gibberellins is to provide the chemical cue for seeds to germinate and grow. When conditions are right for a seed to germinate—when rising temperatures begin to melt frost, for example—these growth-promoting hormones give the green light for a seedling to grow.

Applying gibberellins to a young plant can increase the length of its stem, which also indirectly increases the size of its fruits—a fact that makes them very useful in agriculture. Gibberellins are commonly used to increase the size of seedless grapes, for example. On an untreated seedless grape plant, the stem remains relatively short, so the bunches of grapes growing on the stem are clustered densely together, resulting in small grapes. When sprayed with gibberellins, the stems grow longer, giving the grapes more room to grow. Because seeds are the normal source of gibberellins, seedless grapes have no source of gibberellins to help them grow naturally, which is why farmers need to spray them in the first place.

GIBBERELLINS
Plant hormones that cause stem elongation and cell division.

INFOGRAPHIC 30.9

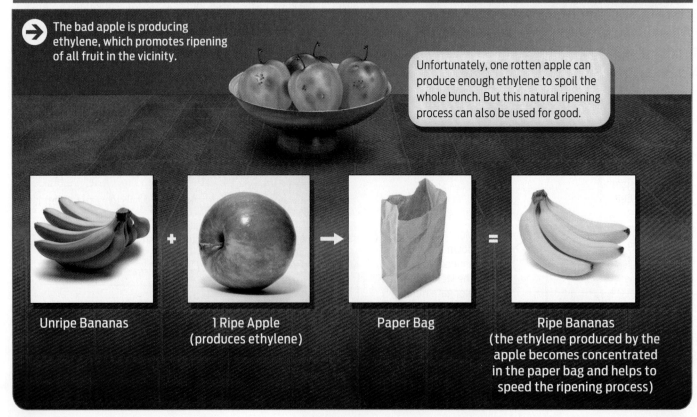

One Bad Apple Does Spoil the Bunch!

→ The bad apple is producing ethylene, which promotes ripening of all fruit in the vicinity.

Unfortunately, one rotten apple can produce enough ethylene to spoil the whole bunch. But this natural ripening process can also be used for good.

Unripe Bananas + 1 Ripe Apple (produces ethylene) → Paper Bag = Ripe Bananas (the ethylene produced by the apple becomes concentrated in the paper bag and helps to speed the ripening process)

Gibberellins play many roles in modern agriculture. Some genetically modified crop plants are *less* sensitive to gibberellins and as a result produce dwarf varieties. Dwarf wheat plants, for example, produce shorter stems and remain small. Yet when treated with fertilizer to encourage growth by providing important nutrients, these dwarf plants devote more of their energy and resources to making wheat grains rather than growing tall stems, greatly improving the yield of wheat grains relative to the less valuable stems (**Infographic 30.10**).

In many species of plant, seed germination is controlled by the balance between the growth-promoting properties of gibberellins and the growth-inhibiting effects of another hormone, **abscisic acid (ABA).** You can think of ABA as the brakes and gibberellins as the gas pedal. Heavy spring rainfalls dilute the levels of ABA in a seed, allowing the gibberellins to step on the gas, ushering in a new season's blooms.

INFOGRAPHIC 30.10

Hormones Trigger Plant Growth and Development

 Certain plant hormones regulate growth and development of plants. When humans apply these hormones to crops, plant growth can be dramatically enhanced.

Gibberellin hormones enhance plant growth:

This pumpkin and the larger bunch of grapes on the right were treated with gibberellin hormones. Spraying plants with gibberellins enhances the number and size of the fruit produced.

The relative amounts of ABA and gibberellin trigger germination:

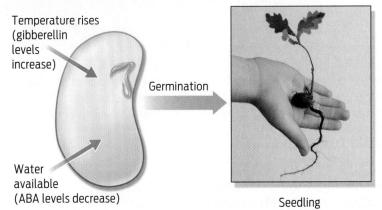

Temperature rises (gibberellin levels increase)

Germination

Water available (ABA levels decrease)

Seedling

Gibberellins are produced when temperatures rise. ABA is diluted by spring rains. The presence of more growth hormone and the absence of inhibitory ABA result in germination and seedling growth.

Why are some plants poisonous?

A From juicy peaches and succulent strawberries to zesty basil and peppery arugula, many plants are incontestably delicious. Humans aren't the only ones who think so. Many herbivores such as insects, birds, and rodents find plant parts tasty and irresistible. This is both helpful and hurtful to a plant. On the one hand, plants rely on animals to eat their fruits and disperse their seeds. On the other hand, plants must ensure that only noncrucial parts of the plants are eaten by other organisms. Eating a plant's fruit is one thing; eating all of its leaves is quite another.

Over the course of their evolution, plants have adapted in many ways to protect their important parts from an herbivore's chomping. Some defenses are mechanical: the stems of a raspberry plant are covered in prickly spines to prevent unwanted chewing; holly leaves are waxy and difficult for insect jaws to grasp. Other defenses are chemical: leaves of the tobacco plant produce nicotine, which is toxic to insects; the bark of the South American evergreen cinchona tree produces quinine, an extremely bitter substance that many animals find distasteful (except certain humans, who use it in their gin and tonics). Such antiherbivory chemicals are a highly effective way to deter pests from eating a plant's leaves.

While a plant's fruits are often tasty and meant to be digested, seeds generally are not. The seed contains a new plant, and therefore it must be protected. Many seeds are encased within an indigestible shell that prevents them from being destroyed by an animal's stomach juices. The unlucky animal that succeeds in breaking open the shell and eating the seed is in for an unwelcome surprise. Seeds are sources of some of the most potent poisons on earth, including ricin, cyanide, and strychnine. Ricin, found in castor beans, can be lethal to animals in quantities as little as two beans. Cyanide, which is found in small doses in the seeds of peaches, apricots, and apples, kills by interrupting cellular respiration in mitochondria; unable to make ample amounts of the short-term energy-storage molecule ATP, nerve and muscle tissues quickly shut down (Infographic 30.11).

Seeds are sources of some of the most potent poisons on earth, including ricin, cyanide, and strychnine.

Plants also use chemicals to combat other plants. Blue gum eucalyptus trees (*Eucalyptus globulus*), for example, secrete a sticky gumlike substance that acts as a deterrent to the germination and growth of noneucalyptus plants in the nearby vicinity.

Perhaps the most interesting method of deterrence that plants use is a kind of unwitting alliance. Some plants, such as wild tobacco, emit potent vapors when they are eaten by insect pests. The vapors, in turn, attract natural predators of the insects, which are thus enlisted in the plant's defense. This "enemy of my enemy" approach is an example of mutualism (see Chapter 22).

Individual plants can even communicate with other individuals of their species and unite against a common enemy. At the first sign of grazing by a hungry giraffe or elephant, for example, African acacia trees release a bad-tasting poison into their leaves. At the same time, they release a gaseous chemical—the versatile hormone ethylene—which drifts out of the stomata of their leaves. Other acacias in a 50-yard radius detect the gas and are prompted to start releasing poison in their leaves, too.

Although antiherbivory chemicals complicate an herbivore's life, they are often quite useful to

Plants Defend Themselves

Physical Defenses:

Raspberry thorns can impale hungry insects.

The jaws of insects that try to feed on holly leaves may not be strong enough to penetrate their waxy coat.

Chemical Defenses:

The leaves of the tobacco plant contain nicotine which is toxic to insects.

The bark of the cinchona tree contains the bitter chemical quinine, which deters insects from feasting on it.

Peach pits have a tough exterior that protects the embryo inside. If an animal is successful in cracking the seed open, it is met with lethal toxins like cyanide.

When a hornworm feeds on a wild tobacco plant, the plant releases chemicals that both repel the worm and attract the worm's predators.

humans. Some of our most important medicines, in fact, are extracts of plant chemical defenses, including aspirin, morphine, digitalis, and the anticancer drug paclitaxel (its proprietary name is Taxol), which was originally obtained from the bark of the Pacific yew tree. ∎

▶ Summary

■ Nearly all plants share the same basic structure: an aboveground shoot system that includes the stem and photosynthetic leaves, and an underground root system that anchors the plant in the soil and absorbs water and nutrients.

■ All plant cells are surrounded by a cell wall made of the complex carbohydrate cellulose. A central water-filled vacuole creates turgor pressure against the cell wall and helps a plant stand up.

■ Some plant tissues have cells with an additional inner cell wall made of cellulose and lignin. Lignin is extremely durable and is what makes plant tissues woody.

■ Plants have a vascular system. Xylem transports water and dissolved nutrients from the roots to the shoots. Phloem transports dissolved sugars throughout the plant body.

■ Water transport is powered by evaporation through stomata in leaves, a process known as transpiration. Plants have mechanisms to control the amount of water lost by transpiration.

■ Plants undergo primary (vertical) and secondary (horizontal) growth. Secondary growth is mostly growth of xylem tissue. Tree rings reflect the growth of xylem tissue from season to season.

■ Plants are autotrophs, able to make their food from sunlight and carbon dioxide (by photosynthesis). They also require a few additional nutrients, primarily nitrogen. Plants obtain nitrogen from the soil, where bacteria convert it into a usable form. In nutrient-poor soils, some plants obtain nitrogen by trapping and digesting insects.

■ Some plants have adapted to hot, dry climates keeping their stomata closed during the day and by collecting carbon dioxide for photosynthesis primarily at night.

■ A variety of pigments assist with photosynthesis. Chlorophyll is the predominant of these pigments and is responsible for the green color of plant leaves. Other pigments contribute to photosynthesis and are visible in the fall, after chlorophyll has broken down.

■ Many plants reproduce sexually and have male and female sexual organs that produce haploid gametes. Plants can be either unisexual or hermaphroditic.

■ Some plants rely on pollinators such as insects to deliver male pollen to female reproductive structures. Other plants use wind to deliver pollen to the female structures.

■ Seed-bearing plants (gymnosperms and angiosperms) produce embryonic plants encased within a protective seed that can disperse a great distance from the parent plant.

■ Plants respond to their environment by various tropisms. Phototropism is growth toward light; gravitropism is growth in response to gravity; thigmotropism is growth in response to touch or wind.

■ Plants produce hormones that contribute to growth and development: ethylene contributes to fruit ripening; gibberellins and ABA regulate germination and stem growth; auxin controls cell elongation.

■ Plants have physical and chemical defenses to fend off herbivores. Humans make use of these defenses in developing pharmaceuticals.

Plants have unique structures, as well as specific nutritional and reproductive strategies.

➔ KNOW IT

1. Draw and label a diagram of two adjacent plant cells. Include key intracellular structures.

2. Which of the following statements represents a true distinction between xylem and phloem?
   a. Xylem provides support only; phloem provides transport.
   b. Xylem provides water and nutrient transport; phloem provides sugar transport.
   c. Xylem transports materials from shoots to roots; phloem transports materials in either direction.
   d. Xylem transports sugars in either direction; phloem transports water from roots to shoots
   e. all of the above

3. What is the function of the cuticle?
   a. It enables neighboring cells to stick together.
   b. It provides rigidity to the cell wall.
   c. It is toxic to many herbivorous insects.
   d. It prevents water loss.
   e. It is sticky and helps pollen to stick to a plant during pollination.

4. Plants are autotrophs and can make sugar from CO_2. How do they obtain their CO_2?
   a. through stomata
   b. by absorption through the root system
   c. by digesting insects
   d. by breaking down carbon-rich carbohydrates stored in roots
   e. a and b

5. What characterizes a dormant seed?
   a. the presence of sperm
   b. the presence of eggs
   c. the presence of an embryonic plant
   d. the absence of ABA
   e. the presence of several pollen grains

6. Describe how fertilization follows pollination in an angiosperm. What has to happen, and what plant structures are involved?

7. Which of the following pigments are present in green leaves late in the summer?
   a. chlorophyll
   b. carotene
   c. xanthophyll
   d. a and b
   e. all of the above

8. Which hormone helps fruit to ripen?
   a. auxin
   b. ethylene
   c. gibberellins
   d. estrogen
   e. anthocyanin

9. If you wanted a plant to grow really tall, which hormone should you apply?
   a. auxin
   b. ethylene
   c. gibberellins
   d. anthocyanin
   e. ABA

➔ USE IT

10. Paper is made from wood that is broken down to pulp. Why are lignin-digesting enzymes included in the pulping process? Would these enzymes have to be included in the pulping process if paper were made from green leaves? Explain your answer.

11. Describe the "conflict" that plants face with respect to opening and closing their stomata.

12. Are trumpet pitchers strict autotrophs? Explain your answer.

13. If you applied to the soil around your plants a chemical that kills bacteria (but not plants), why might your plants die?

14. Why may the bright coloration of a trumpet pitcher have a different function than the bright colors of yellow or orange squash blossoms? (Think about this: squash are pollinated by bees.)

15. If a plant could not make chlorophyll, would you expect it to survive? Why or why not?

16. Why do seedless grapes need hormone treatment to develop big clusters of big grapes, while seeded varieties can develop large fruits without exogenous (that is, externally applied) hormones?

17. Nopales are cactus pads (the large, thick "leaves" of the prickly pear cactus) and make a delicious salad. What antiherbivory mechanism fails when we succeed in making ensalada de nopales—prickly pear salad?

SCIENCE AND ETHICS

18. Paclitaxel (Taxol) is a cancer chemotherapy drug that was initially discovered in the bark of the Pacific yew tree (*Taxus brevifolia*). This species is very slow growing, and occurs only in a small region of the Pacific Northwest. Discuss the conflicts between human needs (for example, anticancer drugs) and the impact of those needs on sensitive plant populations. What are possible solutions to these tensions?

Answers

Chapter 25

1. Anatomy is the study of the structure of organisms, for example the structure of the human body and its "parts." Physiology is the study of how organisms function and maintain homeostasis.

2. (1) mucus-secreting cell of the small intestine (cell); (2) the layer of muscle that contributes to the function of the small intestine (tissue) (3) small intestine (organ); (4) digestive system (organ system)

3. a

4. Physiology. Diet influences a number of physiological processes (for example, levels of blood sugar, glycogen storage, energy balance), as does exercise (for example, thermoregulation, energy expenditure).

5. Because the heart is composed of a number of tissue types, including nervous tissue and muscle.

6. Homeostasis is the ability to maintain a relatively constant internal environment, even when the external environment changes.

7. When body temperature drops, one response is to generate heat through shivering. This requires that glucose be available to the muscle, so that the muscle has an energy source for its contraction during shivering. Glucagon is a hormone that promotes the breakdown of glycogen to glucose, providing fuel for muscles.

8. Cold conditions produce a *low body temperature. Sensors* can detect this drop in body temperature and send a signal to the *hypothalamus.* The hypothalamus in turn sends a signal to *muscles,* which act as *effectors* in this case–contracting by shivering to produce heat and bring about a *normal body temperature.*

9. c

10. The hypothalamus receives input about body temperature from sensors and sends out instructions to effectors. If the hypothalamus is damaged, then it may not receive the information that the core temperature is dropping, or it may not be able to send information to muscles to instruct them to shiver.

11. High insulin is the result of high blood sugar. High blood sugar triggers the pancreas to release insulin. In the presence of insulin, many cells are signaled to take up glucose from the blood. This causes a drop in blood-sugar levels. Now that blood sugar is lower, there is no longer a signal to the pancreas to release insulin, so insulin levels drop as well.

12. Sweating: as sweat evaporates, it cools the surface (in this case, the skin) from which it is evaporating. Vasodilation: by increasing blood flow to the skin, more heat can be released to the environment.

13. In the short term, individual Tibetan Sherpas have acclimatized to the relatively hypoxic conditions on Mount Everest: for example, they likely have relatively large numbers of oxygen-transporting red blood cells in their circulation. In the long term, the Sherpa population has evolved to have high fitness in the low-oxygen environment at high altitude: for example, some alleles encoding hemoglobin may be more efficient at binding oxygen when it is present only at low levels, and these alleles would have been selected for over generations.

14. Research into how the body functions can contribute to understanding of what happens during the course of a variety of diseases, as well as suggesting possible treatments. For example, the study of endocrinology can lead to a better understanding of diabetes (characterized by high blood sugar), stress (many of the hormones produced by the adrenal glands are associated with stress), and cardiovascular disease (heart and blood vessel function).

15. Answers regarding financial responsibility may vary. Preparation for unexpected weather (in a winter climate) should include sufficient clothing to stay warm (prevention of hypothermia), sufficient food to provide energy for the hike and more (in case of being lost or stranded, or needing to fuel shivering muscles), adequate water (to prevent dehydration). If the hike is in an extreme environment, adequate preparation would include sufficient acclimatization to that environment (for example, time to acclimatize to low oxygen pressure).

    In a climate that may become extremely hot, adequate water to stay hydrated is critical, especially as water will be lost by evaporation (sweating) as the body tries to remain cool. Adequate food will also be important. In this environment, severe sunburn can be as dangerous as severe frostbite in cold climates, so adequate sun protection is essential.

Chapter 26

1. (1) mouth; (2) esophagus; (3) stomach; (4) small intestine; (5) large intestine

2. d

3. An expandable stomach allows you to eat more than you can immediately process–the stomach can expand to accommodate the extra food and hold on to it until it can be processed through the small intestine. Without the expandable stomach, you would have to continuously eat small amounts of food.

4. The gallbladder, liver, and pancreas are all accessory organs that secrete enzymes and other substances into the digestive tract. They are not part of the main tube of the digestive tract, which includes the mouth, stomach, and small intestine.

5. a

6. e

7. a

8. They are both digestive enzymes.

9. pancreas

10. a

11. Both are involved in processing fats, but they do so in distinct ways. Bile salts help emulsify large globules of fat to small droplets. These droplets can then be acted on by lipase, which chemically digests triglycerides.

12. The duct between the pancreas and the small intestine allows digestive enzymes produced in the pancreas to enter the small intestine, where they will digest macromolecules in food. When the duct is blocked, the digestive enzymes remain in the pancreas, where they will digest (and damage) the pancreatic tissue.

13. Yes. Food will still pass from your small intestine into your large intestine because food is propelled through your digestive tract by muscular contractions, not gravity.

14. Gallbladder removal leads to the inability to store bile salts. This in turn means that fats cannot be fully processed. Alli is a lipase inhibitor, so taking Alli means that fats cannot be digested. Because undigested fats cannot be absorbed, they are eliminated as fats from the digestive tract, contributing to greasy and loose stools.

15. Someone with a high BMI can reduce the number of Calories consumed and work to burn more Calories through exercise. Some considerations about financing weight-loss surgery and treating obesity-related diseases include the cost of healthful food compared to the cost of highly processed junk food and compulsive behavior around food as well as other behavioral factors that influence food consumption.

16. The surgery is not without risks and has a long recuperation time. Major changes to eating habits after the surgery are necessary to avoid serious discomfort and to lose weight. Not everyone loses the same amount of weight, and some people regain lost weight. Someone at high risk for obesity-related diseases should considering this surgery in consultation with a physician and surgeon. Someone not yet in the very high risk zone in terms of BMI should discuss with a physician the relative risks and benefits of the surgery and the options for weight-loss interventions that do not require surgery.

Chapter 27

1. a light-detecting receptor in the eye–PNS; the amygdala–CNS; a pain receptor in the skin–PNS; the spinal cord–CNS; the thalamus–CNS

2. d

3. b

4. movement: cerebellum; body temperature: diencephalon (specifically the hypothalamus)

5. It is two way. Information from the environment comes in toward the brain, and instructions from the brain go out to the appropriate effectors.

6. Muscles are effectors and receive signals to contract from effector neurons. If these effector neurons lose their myelin sheath, the signal doesn't reach the axon terminal, and therefore the information is not communicated to the muscle. The muscle is capable of contracting but is not receiving the appropriate signals to do so.

7. d

8. b

9. c

10. It diffuses across the synapse and binds to receptors on the membrane of the next cell in the signaling pathway (the postsynaptic cell), which could be another neuron or an effector such as a muscle or gland.

11. Electrical signaling occurs along the length of the axon as characteristic patterns of ion flow, that is, action potentials. Chemical signaling occurs by the release of neurotransmitters (chemical signaling molecules) from the axon terminals of neurons. The neurotransmitters cross the synapse and bind to receptors on the receiving cell.

12. If dopamine remains in the synapse, there will effectively be more dopamine present, which will trigger feelings of pleasure.

13. As acetylcholine normally signals muscles to contract, and Botox prevents the release of acetylcholine into synapses, the muscles will not receive signals to contract, and will be paralyzed in a relaxed state.

14. Because multiple sclerosis is a demyelinating disease, the action potentials in the motor neurons will not reach the axon terminal. Therefore, there will be less acetylcholine released. This explains why those with multiple sclerosis experience muscle weakness: their muscles are not getting the neurotransmitter signal to contract.

15. b

16. In the presence of the drug, dopamine levels are elevated, and cells are bombarded with dopamine. The response of these cells is to down-regulate their dopamine receptors (so that there are fewer active receptors to respond to dopamine). With fewer receptors, there has to be a stronger stimulus (that is, more drug) to achieve equivalent feelings of pleasure.

17. Probably sadder. Even if they produced the same amount of dopamine as others, with fewer dopamine receptors they have an overall reduced response to dopamine. As dopamine is involved in feelings of pleasure, those who respond less well to dopamine will likely have lower levels of pleasure, and hence may feel sadder than others.

18. The loss of dopamine-producing neurons means that there will be less dopamine present. This will result in lower levels of happiness or pleasure, seriously affecting someone with Parkinson disease.

19. This difference is likely to result from politics and social conventions. Cigarette smoking has long been a socially accepted norm (even if not a healthy one), while cocaine is not.

20. There are many possible responses. Some considerations are: not all people with addictions are "junkies" or morally inferior; addictions can have serious health and behavioral consequences, involving treatments that can be costly to the taxpaying public; by understanding the causes of addiction, we may be able to contribute to reducing the need for these treatments; much of the research is basic research that contributes to our fundamental understanding of the nervous system, which can lead to treatments for many neurological diseases, such as Parkinson disease and Alzheimer disease; research can also benefit people who suffer from neurological traumas, include posttraumatic stress disorder and severe nervous system injury.

Chapter 28

1. c
2. b
3. a: The uterus is the organ in which embryos and fetuses develop. The cervix is the opening between the cervix and the vagina. b: The endometrium is the lining of the uterus. The embryo implants into the endometrium.
4. seminiferous tubules→epididymis→vas deferens→urethra
5. One of the risks is inflammation that can lead to permanent scarring of the oviducts. The inflammation can be silent, in that there are no symptoms. Similarly, scarred oviducts present no problems until a woman tries to become pregnant. Then she may encounter infertility, as the scarring prevents egg and sperm from meeting.
6. luteinizing hormone (LH) (anterior pituitary); follicle-stimulating hormone (FSH) (anterior pituitary); testosterone (testis); estrogen (ovary); progesterone (ovary); hCG (embryo)
7. a
8. e
9. FSH and LH are not produced until puberty (when females begin to experience menstrual cycles). Thus, women of reproductive age have substantially higher levels of FSH and LH than girls who have not reached sexual maturity.
10. Testosterone is the primary hormone responsible for sperm development in males and would have to be targeted in order to prevent sperm development. However, blocking testosterone production or activity would also cause a loss of libido and loss of many masculine characteristics, side effects that would not be tolerable.
11. b
12. Both IVF and IUI generally rely on hormonal stimulation of women, so that they will ovulate several mature eggs at the same time. In IVF, those eggs are surgically removed, mixed with sperm in a petri dish, and the resulting embryos placed in the uterus, where it is hoped that they will implant into the endometrium. In IUI, at about the time of ovulation (of possibly multiple eggs), sperm are placed directly into the uterus, in hope that fertilization will occur in the uterus, and that any embryos produced will implant in the endometrium.
13. a: blocked epididymis–this could be detected by imaging; b: polycystic ovary syndrome–this could be diagnosed by a combination of symptoms (for example, irregular menstrual cycles) and and assay of androgen levels (elevated levels of androgens will be found in an affected woman); c: menopause–as menopause is marked by the cessation of menstrual cycles, a woman probably near the age of 50 who has stopped menstruating and whose levels of reproductive hormones are low is probably in menopause; d: oviduct scarring–imaging (for example, ultrasound to visualize an internal blockage) would confirm this diagnosis.
14. In IVF, a fertility specialist can choose how many (or few) embryos to implant. In IUI, there is no way to control the number of eggs that are ovulated in response to the hormonal treatment. Thus, a very large number of eggs could be ovulated, and if all were to be fertilized by the introduced sperm, a large number of embryos could result.
15. Answers will vary. Some of the factors to consider include the health risks to premature babies (and the fact that multiples are often premature); the health care costs (and the question of who bears the burden of those costs); the ability of parents to raise many children of the same age; and the strong desire of infertile couples to be able to have their own children.
16. There are many considerations. In favor of regulation: (1) Clinics wouldn't compete on the basis of success rates, thus reducing the pressure to implant many embryos. All clinics would be operating on a level field. (2) If the number of embryos is regulated, that would reduce the number of multiple pregnancies in which parents may have to consider selective reduction. (3) If regulation reduces the number of multiple births, then the number of premature babies should decline, reducing the burden of illness. Against regulation: (1) Many regulations would reduce the number and rate of successful pregnancies. This places a burden on parents who desperately want to have their own children, as they may have to endure repeated disappointment. (2) If the success rate is reduced, insurance may be less likely to pay for a procedure that has a small chance of success, increasing the financial burden on the patients. (3) Doctors will be unable to pursue all possible options for their patients. (4) Some unscrupulous doctors may set up illegal clinics, and desperate patients may seek treatment in clinics that do not meet general safety standards.

Chapter 29

1. d
2. They must infect the appropriate cell type. Once they enter that cell, they take over the host cell, directing it to replicate the virus.
3. Poliovirus infects and damages nerve cells, which are unable to replicate (and therefore to repair themselves). Influenza infects cells lining the respiratory tract, which are frequently replaced.
4. Antigenic drift is the result of small changes (mutations) in the viral genome. Antigenic shift is the result of segmented viruses exchanging genome segments. This means that a given virus can end up with genes from a completely different virus.
5. Viruses are not made of cells–they must infect host cells in order to replicate. Bacteria are prokaryotic organisms. The vast majority are capable of replicating without entering into and infecting host cells.
6. Each of these viruses has a different host cell. Poliovirus ultimately infects cells of the nervous system, ultimately causing paralysis. Influenza virus infects cells lining the respiratory tract, causing respiratory tract symptoms. HIV

infects cells of the immune system, causing an immune deficiency.

7. skin, which provides protection against pathogen entry; enzymes in tears and saliva that digest components of pathogens; phagocytes that ingest and digest pathogens

8. They are both phagocytic cells.

9. As innate immunity is nonspecific, the innate responses to *E. coli* and to *S. aureus* would be the same. The innate response does not recognize specific pathogens.

10. Neutrophils are critical players in the inflammatory response. Someone with a deficiency of neutrophils would therefore not be able to mount an effective inflammatory response.

11. As influenza can directly and indirectly (through the inflammatory response) damage lung tissue, it destroys the first line of defense against the entry of pathogenic bacteria.

12. Anti-inflammatory drugs suppress the inflammatory response, leaving a person more vulnerable to infections, as bacteria may not be contained or destroyed at the site of entry.

13. Innate immunity is always active and nonspecific. It is present since birth and does not strengthen over time. In contrast, adaptive immunity is specific to a particular pathogen. It is not always "on"–it is induced by the presence of the specific pathogen and strengthens with repeated exposure (because of memory B cells).

14. B cells are lymphocytes that can be activated during an adaptive immune response. Upon activation, B cells specialize into antibody-producing plasma cells, which produce large numbers of antibody molecules.

15. d

16. b

17. a: Yes. b: The memory B cells from a previous exposure will be specific for the particular strain of influenza that was circulating at the time of that exposure. Because of antigenic drift, the influenza virus changes seasonally, so next year's virus will not be identical to the previous virus and this person will need to raise a new primary response against the new flu. c: Because H1Ni arose from antigenic shift, it is substantially different from the seasonally circulating influenza strains. This person's memory B cells will not be specific for this strain of influenza.

18. Antigenic shift results in "mixing and matching" of viral genes from different viruses. This results in a virus that has genes from viruses that typically infect pigs or birds and genes from viruses that typically infect humans. Because the proteins encoded by the pig or bird virus have not been in viruses that infect humans, there is no existing immunity to these essentially completely new proteins, so these "hybrid" viruses can run rampant through the global population.

19. a: Because our own skin cells are recognized as self and (except in cases of autoimmunity) the immune responses are not directed at self. b: Yes. The innate responses are nonspecific. They do not recognize individual pathogens, and are equally effective against all pathogens. c: No. The adaptive response to *Staphylococcus aureus* is specific to *Staphylococcus aureus* and will not be effective against a different bacterium.

20. As helper T cells are critical in mounting an adaptive immune response, the destruction of helper T cells by HIV renders someone who is infected unable to produce an effective adaptive response to pathogens and thus vulnerable to a variety of infections.

21. A person who has been vaccinated has a much lower risk of becoming infected with that pathogen. If everyone is vaccinated, there will be no reservoir of susceptible individuals in whom the pathogen can survive. This can reduce the spread of the pathogen. Also, it will reduce the number of sick people, reducing the burden on health care resources.

22. There are many considerations. One approach is to think about the impact of other disease-preventive conditions (for example, clean water) and what they have done to reduce the burden of disease. Another is to consider a recently developed vaccine (for example, the Heamophilus influenzae type B vaccine for children) and look at the mortality before and after the introduction of this vaccine, keeping in mind that this vaccine was introduced in the era of modern medicine, when there were effective antibiotics and hospital treatments available.

Chapter 30

1. See Infographic 30.1.

2. b

3. d

4. a

5. c

6. Pollination is the delivery of pollen to the female pistil. The pollen grains land on the sticky stigma, and then sperm travel down the style to reach the egg contained in the ovule. Fertilization will generate an embryo that will be contained in a seed.

7. e

8. b

9. c

10. Only wood contains lignified cell walls. Wood pulping requires lignin-digesting enzymes to break down the lignin. Green leaf cells do not have lignified cell walls, so would not require lignin-digesting enzymes to break down.

11. Plants need to open their stomata to let in CO_2 so that they can carry out photosynthesis. Most plants have to do this during the day (when photosynthesis, which is dependent on light, occurs). However, when plants open their stomata to let in CO_2, they lose water by evaporation. In dry climates, this water loss can be substantial and detrimental to the plant.

12. They are not, as they supplement their nutrition with nitrogen from other organisms, specifically insects.

13. Because you might have killed the nitrogen-fixing bacteria in the soil, depriving your plant of fixed nitrogen in the soil.

14. The trumpet pitcher's bright coloration attracts insects that are a source of nitrogen—that is, of "food." Squash blossoms also attract insects (for example, bees), but in this case insects provide "transportation," transferring pollen from the anthers to the insect's body so that the pollen can be carried to the pistil of another plant as the insect makes a tour of blossoms.

15. It would not be able to survive. Chlorophyll is essential for photosynthesis, and without chlorophyll the plant would not be able to make its food.

16. Gibberellins are responsible for the stem elongation that allows large bunches of grapes to develop. These hormones are naturally produced by seeds. Seedless grapes therefore do not produce gibberellins on their own, so must be treated with gibberellins.

17. The prickly spines on the prickly pear cactus are meant to protect the plant from herbivores. By removing the spines for the salad, humans are circumventing this antiherbivory mechanism.

18. Humans have identified many valuable compounds from plants (see Chapter 9). If the plants are fast growing and easy to cultivate, we can grow enough plants to meet our needs. However, if the plants are very slow growing or can't be grown in fields or orchards, then we run the risk of driving the plants to extinction in order to meet our needs. In the long run, this result is not beneficial to either humans or plants. Possible solutions include legislation to protect the plants and their environment, finding alternative sources of the valuable compounds, using alternative strategies to cultivate the plant of interest (for example, grafting), or using chemical synthesis to produce the valuable compound independently of the plant.

Glossary

abscisic acid (ABA) A plant hormone that helps seeds remain dormant.

absorption The uptake of digested food molecules by the epithelial cells lining the small intestine.

action potential An electrical signal within neurons caused by ions moving across the cell membrane.

acclimatization The process of physiologically adjusting to an environmental change over a period of time. Acclimatization is generally reversible.

adaptive immunity A protective response, mediated by lymphocytes, that confers long-lasting immunity against specific pathogens.

allergy A misdirected immune response against environmental substances such as dust, pollen, and foods that causes discomfort in the form of physical symptoms.

amygdala A subregion of the brain that processes emotions, especially fear and anxiety, and is the seat of emotional memories.

anatomy The study of the physical structures that make up an organism.

androgen A class of sex hormones, including testosterone, that is present in higher levels in men and causes male-associated traits like deep voice, growth of facial hair, and defined musculature.

angiosperms Flowering plants.

anterior pituitary gland The gland in the brain that secretes luteinizing hormone (LH) and follicle-stimulating hormone (FSH).

antibody A protein produced by B cells that binds to antigens and either neutralizes them or flags other cells to destroy pathogens.

antigen A specific molecule (or part of a molecule) to which specific antibodies can bind, and against which an adaptive response is mounted.

antigenic drift Changes in viral antigens caused by genetic mutation during normal viral replication.

antigenic shift Changes in antigens that occur when viruses exchange genetic material with other strains.

autoimmune disease A misdirected immune response in which the immune system mistakenly attacks healthy cells.

auxin A plant hormone that causes elongation of cells as one of its effects.

axon The long extension of a neuron that conducts action potentials away from the cell body toward the axon terminal.

axon terminal The tip of an axon, which communicates with the next cell in the pathway.

B cells White blood cells that mature in the bone marrow and produce antibodies during the adaptive immune response.

bile salts Chemicals produced by the liver and stored by the gallbladder that emulsify fats so that they can be chemically digested by enzymes.

brain An organ of the central nervous system that integrates and coordinates virtually all functions of the body.

brain stem The part of the brain that is closest to the spinal cord and which controls vital functions such as heart rate, breathing, and blood pressure.

cell body The part of a neuron that contains most of the cell's organelles, including the nucleus.

cell-mediated immunity The type of adaptive immunity that rids the body of altered (that is, infected or foreign) cells.

cell wall A rigid layer surrounding the cell membrane of some cells, providing shape and structure. In plant cells, the cell wall is made of cellulose, a complex carbohydrate.

central nervous system (CNS) The brain and the spinal cord.

central vacuole A fluid-filled compartment in plant cells that contributes to cell rigidity by exerting turgor pressure against the cell wall.

cerebellum The part of the brain that processes sensory information and is involved in movement, coordination, and balance.

cerebral cortex The outer layer of the cerebrum, the cerebral cortex is involved in many advanced brain functions.

cerebrum The region of the brain that controls intelligence, learning, perception, and emotion.

cervix The opening or "neck" of the uterus, where sperm enter and babies exit.

chlorophyll The dominant pigment in photosynthesis, which makes plants appear green.

chloroplast The organelle in plant cells in which photosynthesis takes place.

chyme The acidic "soup" of partially digested food that leaves the stomach and enters the small intestine.

colon The first and longest portion of the large intestine; the colon plays an important role in water reabsorption.

complement proteins Proteins in blood that help destroy pathogens by coating or puncturing them.

contraception The prevention of pregnancy through physical, surgical, or hormonal methods.

corpus luteum The structure in the ovary that remains after ovulation and secretes progesterone.

cuticle The waxy coating on leaves and stems that prevents water loss.

cytotoxic T cell A type of T cell that destroys altered cells, including virally infected cells.

dendrites Branched extensions from the cell body of a neuron, which receive incoming information.

diencephalon A brain region located between the brain stem and the cerebrum that includes the thalamus and hypothalamus, among other structures, and regulates homeostatic functions like body temperature, hunger, thirst, and the sex drive.

digestion The mechanical and chemical breakdown of food into subunits so that nutrients can be absorbed.

digestive tract The central pathway of the digestive system; a long muscular tube that pushes food between the mouth and the anus.

dopamine A neurotransmitter that is involved in conveying a sense of pleasure in the brain.

duodenum The first portion of the small intestine; the duodenum receives chyme from the stomach and mixes it with digestive secretions from other organs.

ectotherm An animal that relies on environmental sources of heat, such as sunlight, to maintain its body temperature.

effector A cell or tissue that acts to exert a response based on information relayed from a sensor.

elimination The expulsion of undigested matter in the form of stool.

embryo An early stage of development, an embryo is formed when a zygote undergoes cell division.

emulsify To break up large fat globules into small fat droplets that can be more efficiently chemically digested by enzymes.

endocrine system The collection of hormone-secreting glands and organs with hormone-secreting cells.

endometrium The lining of the uterus.

endotherm An animal that can generate body heat internally in order to maintain its body temperature.

epididymis Tubes in which sperm mature and are stored before ejaculation.

epithelial cells Cells that line organs and body cavities; in the digestive tract they sit in direct contact with food and its breakdown products.

esophagus The section of the digestive tract between the mouth and the stomach.

estrogen A female sex hormone produced by the ovaries.

ethylene A gaseous plant hormone that promotes fruit ripening as one of its effects.

feedback loop A pathway that involves input from a sensor, a response via an effector, and detection of the response by the sensor.

fertilization The fusion of an egg and a sperm to form a zygote.

follicle The part of the ovary where eggs mature.

follicle-stimulating hormone (FSH) A hormone secreted by the anterior pituitary gland. In females, FSH triggers eggs to mature at the start of each monthly cycle.

gallbladder An organ that stores bile salts and releases them as needed into the small intestine.

gibberellins Plant hormones that cause stem elongation and cell division.

glial cells Supporting cells of the nervous system.

glucagon A hormone produced by the pancreas that causes an increase in blood sugar.

glycogen An energy-storing carbohydrate found in liver and muscle.

gravitropism The growth of plants in response to gravity. Roots grow downward, with gravity; shoots grow upward, against gravity.

gymnosperms Cone-bearing seed plants.

Helper T cell A type of T cell that helps activate B cells during humoral responses.

hippocampus The region of the brain involved in learning and memory.

histamine A molecule released by damaged tissue and during allergic reactions.

homeostasis The maintenance of a relatively stable internal environment, even when the external environment changes.

hormone A chemical signaling molecule that is released by a cell or gland and travels through the bloodstream to exert an effect on target cells.

human chorionic gonadotropin (hCG) A hormone produced by an early embryo that helps maintain the corpus luteum until the placenta develops.

humoral immunity The type of adaptive immunity that fights infections and other foreign substances in the circulation and lymph fluid.

hypothalamus A master coordinator region of the brain responsible for a number of physiological functions.

hypothermia A drop of body temperature below 35°C (95°F), which causes enzyme malfunction and, eventually, death.

hypoxia A state of low oxygen concentration in the blood.

inflammatory response An innate defense that is activated by local tissue damage.

immune system A system of cells and tissues that acts to defend the body against foreign cells and infectious agents.

immunity The resistance to a given pathogen conferred by the activity of the immune system.

in vitro fertilization (IVF) A form of assisted reproduction in which eggs and sperm are brought together outside the body and the resulting embryos are inserted into a woman's uterus.

inflammatory response An innate defense that is activated by local tissue damage and that acts to recruit phagocytes to the site of infection to destroy and contain invading pathogens.

ingestion The act of taking food into the mouth.

innate immunity Nonspecific defenses, such as physical and chemical barriers and phagocytic cells, that are present from birth and are always active.

insulin A hormone secreted by the pancreas that regulates blood sugar.

interferon Antiviral proteins produced by virally infected cells to help protect adjacent cells from becoming infected.

intrauterine insemination (IUI) A form of assisted reproduction in which sperm are injected directly into a woman's uterus.

kidney An organ involved in osmoregulation, filtration of blood to remove wastes, and production of several important hormones.

large intestine The last organ of the digestive tract, in which remaining water is absorbed and solid stool is formed.

lignin A stiff strengthening agent found in secondary cell walls of plants.

lipase A fat-digesting enzyme active in the small intestine.

liver An organ that aids digestion by producing bile salts that emulsify fats.

luteinizing hormone (LH) A hormone secreted by the anterior pituitary gland. In females, a surge of LH triggers ovulation.

lymph nodes Small organs in the lymphatic system where B and T cells may encounter pathogens.

lymphatic system The organ system that works with the immune system to defend the body by removing toxins and pathogens from the blood, allowing B and T cells to respond to pathogens.

lymphocyte A specialized white blood cell of the immune system.

macrophage A phagocytic cell that resides in tissues and plays an important role in the inflammatory response.

memory cell A long-lived B or T cell that is produced during the primary response and that is rapidly activated in the secondary response.

menstruation The shedding of the uterine lining (the endometrium) that occurs when an embryo does not implant.

motor neurons Neurons that control the contraction of skeletal muscle.

myelin A fatty substance that insulates the axons of neurons and facilitates rapid conduction of action potentials.

natural killer cell A type of white blood cell that acts during the innate immune response to find and destroy virally infected cells and tumor cells.

nerve A bundle of specialized cells that transmit information.

nervous system The collection of organs that sense and respond to information, including the brain, spinal cord, and nerves.

neurons Specialized cells of the nervous system that generate electrical signals in the form of action potentials.

neurotransmitter A chemical signaling molecule released by neurons to communicate with neighboring cells.

neutrophil A phagocytic cell in the circulation that plays an important role in the inflammatory response.

nitrogen fixation The process of converting atmospheric nitrogen into a form that plants can use.

organ A structure made up of different tissue types working together to carry out a common function.

organ system A set of cooperating organs within the body.

osmolarity The concentration of dissolved solutes in blood and other bodily fluids.

osmoregulation Maintenance of a relatively stable volume, pressure, and solute concentration of bodily fluids, especially blood.

ovaries Paired female reproductive organs; the ovaries contain eggs and produce sex hormones.

oviduct The tube connecting an ovary and the uterus in females. Eggs are ovulated into and fertilized within the oviducts.

ovulation The release of an egg from an ovary into the oviduct.

pancreas An organ that secretes the hormones insulin and glucagon, as well as digestive enzymes.

pathogens Infectious agents including certain viruses, bacteria, fungi, and parasites. Many pathogens trigger an immune response.

pepsin A protein-digesting enzyme that is active in the stomach.

peripheral nervous system (PNS) All the nervous tissue outside the central nervous system. The PNS collects sensory information and transmits instructions from the CNS.

peristalsis Coordinated muscular contractions that force food down the digestive tract.

phagocyte A type of white blood cell that engulfs and ingests damaged cells and pathogens.

pheromones Chemical signaling molecules released into the environment in order to signal to other members of the same species.

phloem Plant vascular tissue that transports sugars throughout the plant.

phototropism The growth of the stem of a plant toward light.

physiology The study of the way a living organism's physical parts function.

pistil The female reproductive organ of a flower.

placenta A structure made of fetal and maternal tissues that helps sustain and support the embryo and fetus.

plasma cell An activated B cell that divides rapidly and secretes an abundance of antibodies.

pollen Small, thick-walled structures that contain cells that will develop into sperm.

pollination The transfer of pollen from a male stamen to a female pistil.

primary response The adaptive response mounted the first time a particular antigen is encountered by the immune system.

progesterone A female sex hormone produced by endocrine cells in the ovaries (particularly in the cells of the corpus luteum) that prepares and maintains the uterus for pregnancy.

Rhizobium A genus of nitrogen-fixing bacteria that form symbiotic associations with the roots of legumes.

root hairs Tiny extensions of root cells that increase the surface area of roots to enhance their ability to absorb water and nutrients.

root system The belowground parts of a plant, which anchor it and absorb water and nutrients.

salivary glands Glands that secrete enzymes, including salivary amylase, which digests carbohydrates, into the mouth.

scrotum The sac in which the testes are located.

secondary response The rapid and strong response mounted when a particular antigen is encountered by the immune system subsequent to the first encounter.

seed An embryonic plant contained in a protective structure.

semen The mixture of fluid and sperm that is ejaculated from the penis.

seminiferous tubules Coiled structures that constitute the bulk of the testes and in which sperm develop.

sensor A specialized cell that detects specific sensory input like temperature, pressure, or solute concentration.

sensory neurons Cells that convey information from both inside and outside the body to the CNS.

shoot system The aboveground parts of a plant, including the stem and photosynthetic leaves.

small intestine The organ in which most chemical digestion and absorption of food occurs.

spinal cord A bundle of nerve fibers, contained within the bony spinal column, that transmits information between the brain and the rest of the body.

stamen The male sexual organ of a flower.

stomach An expandable muscular organ that stores, mechanically breaks down, and digests proteins in food.

stomata (singular: stoma) Pores on leaves that permit the exchange of oxygen and carbon dioxide with the air and also allow water loss.

stool Solid waste material eliminated from the digestive tract.

synapse The site of communication between a neuron and another cell; the synapse includes the axon terminal of the communicating neuron, the space between the cells, and the site of reception on the receiving cell.

synaptic cleft The physical space between a neuron and the cell with which it is communicating.

T cells White blood cells that mature in the thymus and can destroy infected cells or stimulate B cells to produce antibodies, depending on the type of T cell.

taproot A long, straight root produced by some plants to store water and carbohydrates.

testes (singular: testis) Paired male reproductive organs, which contain sperm and produce androgens (primarily testosterone).

testosterone The primary male sex hormone, which stimulates the development of masculine features and plays a key role in sperm development.

thermoregulation The maintenance of a relatively stable internal body temperature.

thigmotropism The response of plants to touch and wind.

thymus The organ in which T cells mature.

tissue An organized collection of a single cell type working to carry out a specific function.

tongue A muscular organ in the mouth that aids swallowing.

transpiration The loss of water from plants by evaporation, which powers the transport of water and nutrients through a plant's vascular system.

turgor pressure The pressure exerted by the water-filled central vacuole against the plant cell wall, giving a stem its rigidity.

urethra The passageway through the penis, shared by the reproductive and urinary tracts.

uterus The muscular organ in females in which a fetus develops.

vaccine A preparation of killed or weakened microorganisms or viruses that is given to people or animals to generate a memory immune response.

vagina The first part of the female reproductive tract, extending up to the cervix; also known as the birth canal.

vas deferens Paired tubes that carry sperm from the testes to the urethra.

vascular system Tube-shaped vessels and tissues that transport nutrients throughout an organism's body.

vasoconstriction The reduction in diameter of blood vessels, which helps to retain heat.

vasodilation The expansion in diameter of blood vessels, which helps to release heat.

villi (singular: villus) Fingerlike projections of folds in the lining of the small intestine that are responsible for most nutrient and water absorption.

virus An acellular infectious particle consisting of nucleic acid surrounded by a protein shell.

wood Secondary xylem tissue found in the stem of a plant.

xylem Plant vascular tissue that transports water from the roots to the shoots.

zygote A fertilized egg.

Photo Credits

Chapter 25

p. 493: Galen Rowell/Corbis. **p. 495:** Painted Sky Images/SuperStock. **p. 499:** Courtesy of www.alanarnette.com. **p. 500:** (L) Ken Kamler; (R) Galen Rowell/Corbis. **p. 508:** Devendra M. Singh/AFP/Getty Images. **p. 509:** *Insulation Helps Keep Some Endotherms Warm* (T) Gerrit Vyn, (B) Dennis Scott/Corbis. **p. 510:** (T) jimbo/FeaturePics; (B) Clement Philippe/AgeFotostock. **p. 511:** *Some Organisms Generate Heat from Nonshivering Thermogenesis* (T) AfriPics.com/Alamy, (C) Michael Lynch/Alamy, (BL and BR) Dr. Gladden Willis/Visuals Unlimited/Getty Images.

Chapter 26

p. 515: AP Photo/The Daily Reflector, Rhett Butler. **p. 517:** (T) Pete Saloutos/Corbis; (B) Courtesy of Amy Jo Smith. **p. 520:** AP Photo/The Daily Reflector, Rhett Butler. **p. 523:** *Infographic 26.5* Biophoto Associates/Photo Researchers. **p. 524:** *Infographic 26.6* Courtesy of Allergan Medical. **p. 527:** Science Photo Library/Alamy. **p. 528:** Courtesy of Amy Jo Smith. **p. 529:** *Fungi Digest Food Externally* (L) Jack Bostrack/Visuals Unlimited, (R) Biophoto Associates/Photo Researchers. **p. 530:** *Sea Anemones Have an Incomplete Digestive Tract* http://simple.wikipedia.org/wiki/File:Anemone_monterey_madrabbit.jpg [public domain], via Wikimedia Commons from Wikimedia Commons. **p. 531:** *Plants Have No Digestive Tract* (L) Lessadar/FeaturePics, (R) Nancy Nehring/iStockphoto.

Chapter 27

p. 535: Tyler Sipe/The New York Times/Redux. **p. 535:** Dontcut/Dreamstime.com. **p. 538:** (L) AP Photo/The Lexington Herald-Leader, David Perry; (R) Tyler Sipe/The New York Times/Redux. **p. 541:** *Infographic 27.3* Gary Carlson/Photo Researchers. **p. 544:** Image by Lou Beach. **p. 551:** *Jellyfish Have a Nerve Net* (L) Courtesy of Peter A. V. Anderson, The Whitney Laboratory for Marine Bioscience, (R) Lesya Castillo/Featurepics; *Flatworms Have a Primitive Brain* M. I. (Spike) Walker/Alamy.

Chapter 28

p. 557: Jason Winslow/Splash News/Newscom. **p. 559:** Eyevine/Polaris. **p. 560:** Adel Al-Masry/AFP/Getty Images/Newscom. **p. 563:** *Infographic 28.3* David M. Phillips/Photo Researchers. **p. 568:** (from top) Photomac/FeaturePics.com; (L) Superstock, (R) Jenny Swanson/iStockphoto; (L) Tina Sbrigato/iStockphoto, (R) Moodboard/SuperStock; Daniel Garcia/Dreamstime.com; Imagebroker/Alamy. **p. 569:** *Infographic 28.7* Richard Kessel/Visuals Unlimited. **p. 571:** *Infographic 28.9* UHB Trust/Getty Images. **p. 572:** (T) Jason Winslow/Splash News/Newscom, (B) Jeff Steinberg/Matt Smith, PacificCoastNews/Newscom. **p. 575:** *Birds Reproduce Sexually and Fertilize Eggs Internally* James Urbach/Alamy. **p. 576:** *Fish Reproduce Sexually and Fertilize Eggs Externally* (TL) Mark Conlin/Alamy, (inset) Natural Visions/Alamy, (TR) Mark Conlin/Alamy, (B) Genevieve Anderson. **p. 577:** *Bacteria Reproduce Asexually* Dennis Kunkel/Visuals Unlimited.

p. 578: *Yeast Reproduce Both Sexually and Asexually* (T) J. Forsdyke/Gene Cox/Photo Researchers, (C) Dr. George J. Wong, University of Hawaii at Manoa, (B) David Scharf/Photolibrary.

Chapter 29

p. 583: Science Vu/CDC/Visuals Unlimited. **p. 585:** Time Life Pictures/Getty Images. **p. 587:** Courtesy of the National Museum of Health and Medicine, Armed Forces Institute of Pathology, Washington, D.C. **p. 588:** *Infographic 29.2* Row 1: (L), Science Vu/CDC/Visuals Unlimited, (C) Eye of Science/Photo Researchers, (R) George Musil/Getty Images; Row 2: (L) Dennis Kunkel Microscopy/Visuals Unlimited, (C) James Cavallini/Photo Researchers, (R) Eye of Science/Photo Researchers; Row 3 (L) Dennis Kunkel Microscopy/Visuals Unlimited/Corbis, (C) Masamichi Aikawa, M.D./Phototake/Alamy, (R) Original image by Arturo Gonzalez, CINVESTAV, Mexico. **p. 590:** Karen Kasmauski/National Geographic Stock. **p. 591:** *Infographic 29.4* (T) Carolina Biological Supply Company/Phototake, (BL) Cecil H. Fox/Photo Researchers, (BC) Ocean/Corbis, (BR) Evan Kafka/Getty Images. **p. 592:** *Infographic 29.5* (TL) CDC/ Dr. Thomas Hooten, (TR) Carolina Biological Supply Company/Visuals Unlimited, (BL) temet/iStockphoto, (BR) Ed Reschke/Photolibrary. **p. 594:** Michigan Tech Archives. **p. 601:** *Sea Sponges Employ Physical and Chemical Defenses* (L) Dennis Sabo/iStockphoto, (R) Courtesy of Kate Hendry. **p. 602:** *Bacteria Use Enzymes to Defend against Infection* http://en.wikipedia.org/wiki/File:Phage.jpg. **p. 603:** *Sea Stars Have Innate Cellular Defenses* Paul Kay/Photolibrary; *Fruit Flies Have Multiple Physical, Chemical, and Cellular Defenses* (from top) playTOME, FeaturePics.com, Carl-Johan Zettervall and Dan Hultmark, Umeå University, Willott, E., Tran, H. Q. 2002. Zinc and Manduca sexta hemocyte functions. 9 pp. *Journal of Insect Science*, 2.6. Available online: insectscience.org/2.6. Photo by H. Q. Tran.

Chapter 30

p. 607: Pixtal Images/Photolibrary. **p. 608:** Helene Schmitz. **p. 610:** *Infographic 30.1* (T) Nuridsany et Perennou/Photo Researchers, (BL) Perennou Nuridsany/Photo Researchers, (BR) Krilt/Dreamstime.com. **p. 611:** *Infographic 30.2* Steve Gschmeissner/Photolibrary. **p. 612:** GYRO Photography/amanaimagesRF/Getty Images. **p. 614:** *Infographic 30.3* (from left) Pixtal Images/Photolibrary, Courtesy of Chris Moody; Courtesy of Frank Dazzo, Center for Microbial Ecology, Michigan State University; Dr. Jeremy Burgess/Photo Researchers. **p. 615:** *Infographic 30.4* (TL and BL) Dr. Jeremy Burgess/SPL/Photo Researchers, (R) EastEggImages/Alamy. **p. 616:** *Infographic 30.5* (L) Jens Stolt/iStockphoto, (inset) sovlanik/iStockphoto, (R) Don Johnston/Alamy, (inset) Natallia Khlapushyna/Dreamstime.com. **p. 617:** Image Plan/Corbis. **p. 618:** *Infographic 30.6* (L) Chas53/FeaturePics, (R) Rolf Nussbaumer Photography/Alamy. **p. 619:** *Infographic 30.7* (TL) Ray Roper/iStockphoto, (TC) cross section, Olga Demchishina/iStockphoto, (TR) Shaun Pimlott/iStockphoto, (B, from left) Stephen Dalton/Getty Images, Suijo/Dreamstime.com, Mark A. Johnson/Alamy, Juniors Bildarchiv/Alamy,

Index

Note: page numbers followed by f indicate figures; those followed by t indicate tables.